BSA 350, 500 and 600 Pre-unit Singles Owners Workshop Manual

by Mansur Darlington

Models covered

348cc	B31 ohv single	1954 - 1959	499cc	B33 ohv single	1954 - 1960
348cc	B32 Competition model	1955 - 1957	499cc	B34 Competition model	1955 - 1957
348cc	CB and DB 32 Gold Star	1954 - 1957	499cc	CB, DB and DBD 34 Gold Star	1954 - 1961
496cc	M20 side-valve single	1954 - 1955	591cc	M21 side-valve single	1954 - 1961
499cc	M33 ohv single	1954 - 1957			

ISBN 978 0 85696 326 1

Haynes Group Limited
Haynes North America, Inc

www.haynes.com

Acknowledgements

Our thanks are due to Rod Grainger for the loan of his BSA B31 350 cc model that was used for the photographs featured in this Manual. We would also like to thank Lewis and Sons Ltd, Weybridge, Surrey and Russell Motors Ltd, of Battersea, who supplied the spare parts required in rebuilding the B31, and Eric Raymond of Templecombe, Somerset and Brian Hoppe of Lelliot and Bird Ltd, Sherborne, Dorset who kindly loaned many BSA publications.

Leon Martindale (Member of the Master Photographers Association) arranged and took the photographs, Brian Horsfall and Martin Penny assisted with the rebuild.

We would also like to acknowledge the help of the Avon Rubber Company, who kindly supplied the illustrations relating to tyre fitting. Amal Ltd for their carburettor illustrations and NGK Spark Plugs (UK) Ltd for advice about spark plug conditions.

Particular thanks go to Les Thomas of Yeovil who supplied us with the BSA Gold Star featured on the cover.

The author would like to thank Jeff Clew who edited the text and contributed the introduction to the marque on page 5, and Peter Shoemark who carried out the unenviable task of writing captions to all the line drawings, and who also assisted in preparing the manuscript for publication.

About this manual

The author of this manual has the conviction that the only way in which a meaningful and easy to follow text can be written is first to do the work himself, under conditions similar to those found in the average household. As a result, the hands seen in the photographs are those of the author.

Although the BSA 350, 500 and 600 cc models are no longer in production, an example of the B31 model that had covered a considerable mileage was deliberately selected so that the conditions encountered during the strip down and rebuild would be typical of those encountered by the average owner/rider. Unless specially mentioned and therefore considered essential, BSA service tools have not been used. There is invariably alternative means of loosening or slackening some vital component, when service tools are not available and risk of damage is to be avoided at all costs.

Each of the eight chapters is divided into numbered sections. Within each section are numbered paragraphs. Cross-reference throughout this manual is quite straightforward and logical. When reference is made 'See Section 6.10, it means Section 6, paragraph 10 in the same chapter. If another chapter was meant, the reference would read 'See Chapter 2, Section 6.10.

All photographs are captioned with a section/paragraph number to which they relate and are always relevant to the chapter text adjacent.

Figure numbers (usually line illustrations) appear in numerical order, within a given chapter. Fig. 1.1. therefore refers to the first figure in Chapter 1.

Left-hand and right-hand descriptions of the machines and their components refer to the left and right of a given machine, with the rider normally seated.

Motorcycle manufacturers continually make changes to specifications and recommendations, and these, when notified, are incorporated into our manuals at the earliest opportunity.

We take great pride in the accuracy of information given in this manual, but motorcycle manufacturers make alterations and design changes during the production run of a particular motorcycle of which they do not inform us. No liability can be accepted by the authors or publishers for loss, damage or injury caused by any errors in, or omissions from, the information given.

Contents

	Page
Acknowledgements	2
About this manual	2
Modifications to the BSA single cylinder range	5
Introduction	5
Ordering spare parts	6
Routine maintenance	7
Recommended lubricants	12
Working conditions and tools	12
Chapter 1 Engine	14
Chapter 2 Gearbox	58
Chapter 3 Clutch	75
Chapter 4 Fuel system, carburation and lubrication	80
Chapter 5 Ignition system	92
Chapter 6 Frame and forks	101
Chapter 7 Wheels, brakes and tyres	116
Chapter 8 Electrical system	130
Wiring diagrams	140
Conversion factors	142
Safety First!	143
English/American terminology	144
Index	145

BSA Gold Star model

1957 BSA 350 cc B31 model

Modifications to the BSA single cylinder range

During the production life of the BSA single cylinder models many detailed and a smaller number of major modifications were made. Of the detailed modifications, a large proportion relate to the alteration in cycle parts which do not materially affect the methods of overhaul or maintenance. The larger modifications which include improvement in frame design and the utilisation of swinging arm rear suspension, were incorporated on most models by late in 1954. It was in this year also that the Gold Star CB type engine appeared and the Amal 'Monobloc' carburettor succeeded the 'standard' Amal instrument. The objective of this Manual is to cover all models within the range from a time when the major modifications were made. To this end the text has been designed specifically to relate to models after the following engine numbers.

350 cc B31	*Engine No. BB31 - 15001*
	introduced September 1954
350 cc B 32 Competition	*Engine No. BB32A - 301*
	introduced November 1955
350 cc B 32 Gold Star	*Engine No. CB32GS-501*
	introduced September 1954

500 cc M20	*Engine No. BM20 - 2501*
	introduced September 1954
500 cc B 33	*Engine No. BB33-5001*
	introduced September 1954
500 cc M 33	*Engine No. BM33-1301*
	introduced September 1954
500 cc B 34 Competition	*Engine No. BB34A-351*
	introduced November 1955
500 cc B 34 Gold Star	*Engine No. CB34G3-1001*
	introduced 1954
600 cc M 21	*Engine No. BM21-4501*
	introduced September 1954

This manual will also be of considerable help to owners of machines manufactured before these dates. It should be noted, however, that in addition to the standard models produced, a number of machines were supplied to various organisations, and for export, which differed materially from the standard specifications. Because of the very small numbers involved, no mention of these models has been made in the text.

Introduction

The BSA B31 and B33 single cylinder models appeared on the UK market soon after the end of World War 2, the B31 during 1946 when the manufacture of machines for the civilian market was resumed, and the B33 almost a year later. Designed primarily for touring purposes, both models featured an all-iron engine unit and a rigid frame. Shortly after each model had made its debut, a 350 cc and a 500 cc competition variant was also marketed, designated B32 and B34 respectively. With the resumption of motorcycle sport, there was an obvious need for machines of this latter category, which could also be used on the road, if required.

In due course, all the B series models were made available with an optional spring frame of the plunger type, the spring frame being another development in which there was a great deal of interest after the war. Plunger-type rear suspension was, however, only a compromise until something better came along and it was in 1954 that BSA Motor Cycles announced the introduction of an entirely new frame design of the duplex tube type which utilised the swinging arm mode of rear suspension. The new frame went into production during the 1954 season and continued until the B series models were phased out of production during 1961. The use of the new frame necessitated a change in gearbox design, hence it was not possible to convert the older models simply by changing the frame alone.

The 'Gold Star' B32 model made its debut during the 1948 Motor Cycle Show at Earls Court, although this was not really its first appearance. A 'Gold Star' 500 cc model had appeared in the BSA catalogue as far back as 1938, after a similar model, still in the prototype stage, had lapped Brooklands at 107.57 mph in the hands of the late Wal Handley. The war curtailed further development and when the model re-appeared during 1948, certain changes had been made in the specification, including the fitting of telescopic front forks. A 500 cc version was added to the range for the 1949 season. Although the Gold Star models were marketed primarily as high performance road machines, they very soon found application in clubman racing events and engine development continued with this end purpose in mind

until the Gold Star was so outstandingly successful in this type of event that it dominated it completely and brought clubman racing to an end. Development was, however, by no means confined to road racing. Trials and motocross versions of the Gold Star also received their due measure of attention and it was in these other two areas, particularly the latter, that the Gold Star made its mark. Although the 350 cc and 500 cc models were designated B32 and B34 respectively, the words Gold Star were always appended to the designation, to avoid confusion with the standard B32 and B34 competition models. It is a popular misconception that any machine that has either of these latter two model numbers must be a Gold Star. Engine numbers are characterised by the initials GS, interposed between the model designation and the actual engine number, in the case of all genuine Gold Star models. As in the case of the other BSA models, swinging arm rear suspension superseded the older plunger-type rear suspension during the 1954 season.

For those who ran a sidecar outfit, the 500 cc M33 model offered the best prospects, since the frame was equipped with sidecar lugs and the general specification of this model was such that it could be marketed at a slightly lower price than the B33 model it so closely resembled. This model initially had a rigid frame, but was eventually available with plunger-type rear suspension, which it retained until production ceased during 1957. It did not, therefore, have the later-type duplex tube frame.

Last, but certainly not least, mention must be made of the two side-valve models, the M20 and M21, of 500 cc and 600 cc capacity respectively. Although of pre-war origin, both models saw active service during the war, and re-emerged afterwards with certain modifications, such as telescopic front forks, which led to their continued acceptance as an ideal sidecar mount. Like the M33 ohv single, the old type of frame, with plunger rear suspension, was used until the M20 model was discontinued during 1955, and the M21 during 1961. Both models were also available with a rigid frame, as an option.

Ordering spare parts

Although BSA Motor Cycles Limited is no longer in existence many spare parts for the BSA range of pre-unit single cylinder motorcycles are still widely available. A number of motorcycle dealers who acted as BSA agents still retain their remaining stocks of parts. In addition, a number of specialist spare parts dealers now provide a service for owners of these machines. The carburettor and electrical components were used by a large number of manufacturers and as such are often available from specialist dealers catering for entirely different marques.

Always quote the engine and frame numbers in full, especially for these older models. Include any letters before or after the numbers itself. The importance of giving this information cannot be overstressed. Quite fundamental design changes occured when the swinging arm method of rear suspension was adopted and the engine and frame numbers provide the most positive form of correctly identifying the machine concerned. The frame number will be found stamped on the left-hand side of the gusset around the steering head, or on the lower half of the right-hand frame tube. The engine number is stamped on the left-hand crankcase, immediately below the base of the cylinder barrel.

Use only parts of genuine BSA manufacture. Pattern parts are available, but in many instances they will have an adverse effect on performance and/or reliability. Some complete units were available originally on a 'Service Exchange' basis so that costs could be kept to an economic level by the supply of factory-reconditioned replacements. It is worth enquiring whether any of these facilities still exist in the form of left-over stocks on an agent's shelves or the agent's own reconditioned replacements.

Retain any broken or worn parts until a new replacement has been obtained. Often these parts are required as a pattern for identification purposes, a problem that becomes more acute when a machine is classified as obsolete. In an extreme case, where replacements are not available, it may be possible to reclaim the original or to use it as a pattern for having a replacement made. Many older machines are kept on the road in this way, long after a manufacturer's spares have ceased to be available.

Some of the more expendable parts such as spark plugs, bulbs, tyres, oils and greases etc., can be obtained from accessory shops and motor factors, who have convenient opening hours, charge lower prices and can often be found not far from home. It is possible to obtain parts on a Mail Order basis from a number of specialists who advertise regularly in the motor cycle magazines.

Frame number location

Engine number location

Routine maintenance

Periodic routine maintenance is a continuous process that commences immediately the machine is used. It must be carried out at specified mileage recordings or on a calendar date basis if the machine is not used regularly - whichever falls soonest. Maintenance should be regarded as an insurance policy rather than a chore, because it will help keep the machine in peak condition and ensure long, trouble-free service. It has the additional benefit of giving early warning of any faults that may develop and will act as a regular safety check, to the obvious benefit of both rider and machine alike.

The various maintenance tasks are described under their respective mileage and calendar date headings. Accompanying diagrams have been added, where necessary. It should be remembered that the interval between the various maintenance tasks serves only as a guide. As the machine gets older or is used under particularly arduous conditions, it would be advisable to reduce the period between each check.

Each service operation is described in detail. If additional information is required, it will be found under the relevant heading in the appropriate Chapter. No special tools are required for the normal routine maintenance tasks. The tools contained in the kit supplied with every new machine will prove adequate for each task but if they are not available, the tools found in the average household should suffice.

Weekly, or every 200 miles

Checking engine oil level

1   Remove the oil tank filler cap and view the oil level through the orifice. Originally most oil tanks were equipped with an oil level mark transfer. If the mark is no longer visible, an estimate of the oil level height should be made when renewing the oil, by replacing the specified quantity in the tank. If required, replenish the engine oil using the specified grade. Either a monograde or a multi-grade may be used, but avoid mixing the two types of oil.

Battery electrolyte level

Remove the dualseat and detach the battery cover by removing the strap. Check that the electrolyte level in each cell is approximately ¼ inch above the plates. Replenish, if necessary, with distilled water. Do not use tap water which contains many mineral impurities which will reduce the efficiency of the battery.

Rear chain tension and lubrication

Chain adjustment is correct when there is approximately ¾ inch free play in the middle of the lower run. Always test the chain at the highest position in the rotation of the chain. Adjust-

ment is effected by slackening the wheel spindle and torque arm nuts and moving the rear wheel backwards by means of the adjuster bolts in the fork ends. Always move the adjusters an equal amount, to maintain correct wheel alignment. When adjustment is correct, push the wheel hard against the adjuster bolts before tightening the spindle. Do not omit to tighten the torque arm retaining nut.

If there is any doubt about the accuracy of wheel alignment, this can be checked by using a long straight edged plank positioned along the machine. See Fig. 7.10 in Chapter 7.

Lubricate the chain using engine oil or one of the proprietary aerosol lubricants. The latter type is more efficient as it is less prone to being flung off by the fast rotating chain.

Tyre pressures

Check the pressure of both tyres, using a pressure gauge that is known to be accurate. Always check tyres when they are quite cold. After a machine has covered a number of miles, the tyres will warm up, causing the pressure to increase. The correct pressures for each model are as follows:

	Front (psi)	Rear (psi)
M20	17	22
M21	16	18
M33	17	18
B31	17	23
* B32	20	16
B33	17	19
* B34	22	16
* B32 Gold Star	21	22
* B34 Gold Star	21	18

* Tyre pressures for use in competition vary, to suit individual requirements.

Safety check

Inspect each tyre for signs of splitting, cracking or other damage. Use the blade of a screwdriver to remove any small stones or other objects which may be embedded in the treads.

Inspect the machine for loose nuts and bolts or other components, tightening where necessary.

Legal check

Check that all the lights are in working order and that the horn functions.

Monthly, or every 1,000 miles

Complete the maintenance tasks listed under the preceding weekly heading, then the following additional items:

A   The pressed-steel chaincase   B   The aluminium chaincase

a   *Drain screw*
b   *Level screw*
c   *Inspection filler cap*

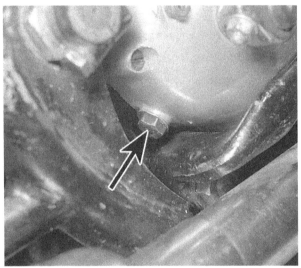

Remove bolt to check gearbox oil level

Oil tank filter is integral with drain plug (B series)

Drain crankcase of oil and ...

... remove sump plate for filter cleaning

Gearbox oil level

Place the machine on level ground and check the gearbox oil content by means of the level screw at the rear of the gearbox. If necessary, remove the oil filler cap which is retained by two screws (threaded cap, M series) and add engine oil until it can be seen emerging from the level orifice. Allow the level to settle, then replace the drain screw and cap.

Primary chain case oil level

Remove the level plug from the chaincase. On pressed steel chaincases the plug is located just below the clutch housing. Aluminium chaincases have the level plug incorporated in the second retaining screw from the front, on the lower run of screws. If necessary, replenish the oil through the filler cap which screws into the top of the chain case. Replace the cap and fit and tighten the drain plug.

General lubrication

Apply a grease gun to the rear suspension plungers, or the swinging arm pivot grease nipples. Grease also the wheel hubs and brake cam spindles, if grease nipples are provided.

Note that both these points should be greased sparingly, to prevent grease from entering the brake drums.

Two months, or every 2,000 miles

Complete all the checks listed under the weekly and monthly headings, then the following items:

Engine oil change

Drain the engine oil from the oiltank and the sump, preferably when the engine is warm so that the oil will flow more easily.

On M series machines, the tank drain plug is in the base of the oil tank. On all other models the drain plug, which incorporates the main oil filter, is fitted into the side of the tank. When draining either oil tank it is worthwhile arranging a cardboard or tin shute down which the oil will flow into the drain pan. Remove the drain plug in the base of the crankcase and then remove the sump plate which is retained by four nuts. The scavenge pump oil filter screen will come away with the plate.

Clean the sump filter, and the main filter, in petrol. On M series machines the filter can be lifted from position after detaching the oil filler cap and removing the filter upper seating.

Replace the sump filter screen and plate together with two new gaskets, one either side of the screen. Refit the main filter and the oil drain plugs.

Replenish the oil tank with either multi-grade 20W/50 engine oil or a monograde engine oil of SAE 30 (winter) or SAE 40 (summer) specifications. The correct quantity is as follows:

M series	5 pints (2.84 litres)
B series	4 pints (2.27 litres)
Gold Star models	5½ pints (3.12 litres)

Primary chaincase oil change

Place a drain pan below the primary chaincase and remove the oil filler cap and the drain plug. On M series models the plug is situated in the base of the chaincase cover. On all other models the drain plug is incorporated in the fourth case retaining screw from the front in the bottom run of screws. Allow plenty of time for the oil to drain, especially on machines fitted with aluminium alloy chaincases. Replace the drain plug or screw, ensuring that the fibre sealing washer is not omitted. Remove the level screw and replenish the casing with engine oil until it begins flowing from the level orifice. Allow the level to settle, then replace the level plug and filler cap.

Primary chain adjustment

Remove the primary chaincase filler plug and turn the engine over by means of the kickstart until the primary chain can be felt to be at its highest point. The chain is in correct tension if a total of ½ inch up and down movement can be felt at the middle of the run. If the tension is not correct, loosen off the gearbox main mounting bolts and, by means of the drawbolt on the right-hand side of the gearbox, move the gearbox forwards or backwards, as required. Tighten the mounting bolts and recheck the tension. Because adjustment of the primary drive chain requires that the gearbox be moved, it follows that after adjustment, the rear chain will also require adjusting. Refer to the relevant heading in the weekly/200 mile routine maintenance schedule.

Rear chain lubrication

Lubrication of the rear chain may be carried out with the chain in situ on the machine, as described under the weekly/200 mile heading. At greater intervals, however, the chain should be removed from the machine, cleaned thoroughly and relubricated by immersion in a hot bath of special chain grease such as 'Linklyfe' or 'Chainguard'. To remove the chain, disconnect it at the spring link. If possible, connect an old chain to one end of the chain, which can then be pulled into place on the gearbox

Move gearbox to adjust primary drive chain

Use proprietary chain lubricant for rear chain

sprocket. By using this method the relubricated chain can be pulled back into position with ease.

On machines equipped with a totally enclosed chaincase, the rear of the chaincase will have to be removed before the chain can be detached. A totally enclosed chain will not require lubrication at such short intervals; every 5,000 miles should prove adequate.

Air filter element

The air filter fitted originally to some models is contained within a box attached to the battery mounting strap by two bolts. To remove the element, pull off the air hose and detach the box from the battery strap. Unscrew the screws retaining the gauze screen at the rear of the box and lift out the filter.

A later, non-standard filter, often fitted after manufacture, is contained in a circular box which screws directly onto the carburettor. This filter is held in position by a single screw. The element of either type of filter should be washed in petrol, allowed to dry and then immersed in thin oil (SAE 20). Allow the element to drain thoroughly, before replacing it in the filter housing.

Valve clearance adjustment

Place the machine on the centre stand so that the rear wheel is clear of the ground. Remove the inspection cover at the bottom of the push rod tube, and unscrew the spark plug. Because the cams incorporate cam quietening ramps, the cam followers (tappets) are only at their lowest point during a very small period in the rotation of each cam. Due to this, valve clearance adjustment must follow a special procedure.

Turn the engine forwards until the inlet valve has just closed. The engine may be rotated by placing the machine in top gear and turning the rear wheel anticlockwise. At this point the engine is in the correct position to check the exhaust valve clearance. Then turn the engine forwards again very slowly until the exhaust valve clearance has just been taken up, but before the valve actually begins to open. The inlet valve clearance may now be set. The gap should be checked by placing a feeler gauge of the appropriate size between the tappet head and the lower end of the pushrod. (Valve stem, on side-valve models).

On the M series, B31 and 33, and the competition models, adjustment is made by slackening the locknut below the tappet head, and screwing the tappet head upwards and downwards on the tappet threads.

The clearances are as follows:

	Inlet	Exhaust
M20 and M21	0.010 in (0.25 mm)	0.012 in (0.30 mm)
B31, 33 and M33	0.003 in (0.08 mm)	0.003 in (0.08 mm)
Competition (B32 and B34)	0.003 in (0.08 mm)	0.003 in (0.08 mm)

Tighten the locknut after adjustment.

Adjustment of the valve clearances on Gold Star models is effected by means of eccentric rocker spindles which may be rotated by slackening the acorn nuts and detaching the two oil feed banjo unions. Place a spanner on the hexagon below the banjo to turn the rocker spindle. The correct valve clearances are as follows:

	Inlet	Exhaust
All models (except touring CB, DB and DBP Gold Stars	0.006 in (0.15 mm)	0.006 in (0.15 mm)
Touring CB and DB Gold Stars	0.008 in (0.20 mm)	0.010 in (0.25 mm)

After adjustment on all models, refit the inspection cover, using a new gasket where required. Note that the tappet clearances must always be checked or re-set when the engine is cold.

Contact breaker gap

Remove the contact breaker assembly cover and turn the engine until the points are in the fully open position. Clean the points faces using a fine swiss file or a strip of emery paper (No. 400) backed by a thin strip of tin. If the points faces are badly blackened, burned or pitted, they should be removed for further attention as described in Chapter 5, Section 5.

Using a 0.012 in (0.30 mm) feeler gauge, check the points gap. This must be carried out with the points in the fully open position or a false reading will result. If the points gap is incorrect loosen the fixed point locknut and screw the points in or out as required. Tighten the locknut and recheck. On alternator models the points gap is adjusted by loosening the fixed point retaining screw and moving it nearer to or further away from the moving point. Tighten the screw and recheck. Lubricate the contact breaker cam before replacing the points cover.

Check tappet clearance with feeler gauge

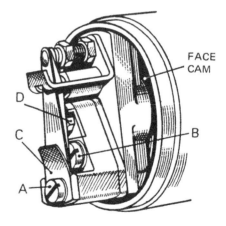

Magneto contact breaker assembly

a   Securing screw
b   Wick holder
c   Backing spring and contact breaker arm
d   Body mounting screw

Spark plug

Remove the spark plug and clean the area around the electrodes with a wire brush and then attend to the points gap with fine emery paper. Set the gap to 0.018 - 0.022 in (0.45 - 0.55 mm) by bending the outer electrode nearer to or further away from the central electrode. Before replacing the plug, lubricate the threads with graphite grease to aid future removal.

Six monthly, or every 5,000 miles

Again complete all the routine maintenance tasks listed previously, then the following additional tasks:

Gearbox oil change

Place a suitable container below the gearbox and remove the drain plug. The gearbox oil must be changed when the oil is warm, to improve the flow. Replenish the gearbox, with oil of the same type used in the engine, after replacing the drain plug. The gearbox capacity is 1 pint. Do not overfill the gearbox or the oil will find its way along the mainshaft into the clutch. Pour the oil in through the filler cap, which is retained by two screws (threaded cap on M series). A cardboard shute aids filling.

Yearly, or every 10,000 miles

After completing the weekly, two monthly and six monthly tasks, continue with the following additional items:

All these items require greater attention than the previously described shorter interval maintenance operations. For full details refer to the relevant Chapters within the main text.

Drain and replenish the fork oil. Check the fork brushes and oil seals for wear, renewing where considered necessary.

Remove the wheels and check the condition of the brake linings. If necessary, renew the lining and/or the shoes.

Remove the wheel bearings and clean thoroughly. Check the bearings and renew if worn. Repack with grease on replacing.

Detach and dismantle the carburettors. Clean all the components and airways, replacing any jets if worn.

Repack the steering head bearings with grease after removal for cleaning.

Remove the cylinder head and give a complete top end overhaul.

Brake adjustment

The interval at which the brakes will require adjustment depends largely on the purpose to which the machine is put and the manner in which it is ridden. It follows that a machine consistently ridden hard will suffer more rapid brake lining wear than one used for commuting only.

Adjustment of the front brake is made by screwing the cable adjuster in or out until a small amount of play is present before the brake comes into operation. The rear brake is of cable or rod operation, depending on the model. The rod brake is adjusted by a knurled nut on the rod. The cable brake is adjusted in a similar manner to that of the front. In both cases the amount of travel of the brake pedal should be approximately two inches before the brake is applied fully.

On 1958-60 B31 and B33 models, an adjustable brake shoe fulcrum pin is fitted to each brake plate, to enable precise adjustment of the brake shoes to be carried out. The procedure for adjusting the fulcrum pin is identical for each wheel. Slacken off the brake cable so that there is plenty of slack. Screw in the fulcrum pin whilst rotating the wheel. When the brake begins to bind, slacken the pin off just sufficiently to allow the wheel to spin freely. If necessary, the cable can then be adjusted to give the correct play at the lever or pedal.

Clutch adjustment

In common with the brakes, clutch wear and the necessity for adjustment depends upon the manner in which the machine is used. Clutch adjustment should always be carried out in two stages, as follows:

Screw in the adjuster at the lower end of the clutch cable so that there is plenty of play in the cable. Remove the gearbox oil filler cap, which is retained by two screws on all models, excluding the M series, where the cap is of the screw type. By means of the screw in the clutch operating shaft adjust the clutch operating arm so that it is parallel with the gearbox casing when the clutch is fully **disengaged**. This gives the optimum amount of leverage and the consequent ease of control. The adjuster screw is fitted with a locknut which must be loosened for adjustment and tightened when adjustment has been completed. The cable should now be adjusted so that there is approximately 1/8 inch play at the the handlebar lever, before clutch disengagement commences. This is of extreme importance in allowing a small amount of play between the pushrod and clutch adjustment screw.

Control cable lubrication

Cable life, and efficiency are dependent on regular lubrication. The intervals at which this should take place are governed by the

Adjust cable by means of adjuster

Do not omit to tighten clutch adjuster locknut

general climatic conditions, but as a general rule, any obvious
grime or stiffness in operation, warrants cleaning and lubrication.

Use motor oil or an all-purpose oil to lubricate the control
cables. A good method for lubricating the cables is shown in the
accompanying illustration, using a plasticine funnel. This method
has a disadvantage in that the cables usually need removing from
the machine. An hydraulic cable oiler which pressurises the
lubricant, overcomes the problem. Do not lubricate nylon lined
cables as the oil will cause the nylon to swell, thereby causing total
cable seizure.

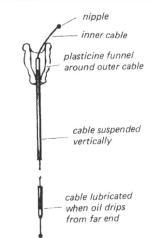

Control cable oiling

Recommended lubricants

	Type	Quantity
Engine	SAE 30 winter SAE 50 summer (or 20W/50)	4 Imp. pints (2.27 litres) or 5½ Imp. pints (3.12 litres) Depending on year of manufacture and model. See Specifications.
Gearbox	Engine oil (or 20W/50)	1 Imp. pint (570 cc)
Chaincase (primary)	Engine oil (or 20W/50)	To level
Front forks	SAE 20 oil	5½ fl. oz (213 cc) (quantity per leg)
Wheel hubs, rear suspension centre stand, saddle nose bolt and all other grease nipples	Multi-purpose grease (Lithium based)	
Rear chain	Graphite grease or aerosol spray	
Contact breaker gap		0.012 in (0.30 mm)
Spark plug gap		0.018 in - 0.020 in (0.45 - 0.55 mm)

Tappet clearance (engine cold)	Inlet	Exhaust	Model
	0.010 in (0.25 mm)	0.012 in (0.30 mm)	M20, M21
	0.003 in (0.08 mm)	0.003 in (0.08 mm)	B31, B33, M33
	0.003 in (0.08 mm)	0.003 in (0.08 mm)	B32, B34
	0.006 in (0.15 mm)	0.006 in (0.15 mm)	All Gold Star, except CB & DB Touring models
	0.008 in (0.20 mm)	0.010 in (0.25 mm)	CB and DB Model Gold Stars (Touring)

Working conditions and tools

When a major overhaul is contemplated, it is important that
a clean, well-lit working space is available, equipped with a
workbench and vice, and with space for laying out or storing the
dismantled assemblies in an orderly manner where they are
unlikely to be disturbed. The use of a good workshop will give
the satisfaction of work done in comfort and without haste,
where there is little chance of the machine being dismantled
and reassembled in anything other than clean surroundings.
Unfortunately, these ideal working conditions are not always
practicable and under these latter circumstances when
improvisation is called for, extra care and time will be needed.

The other essential requirement is a comprehensive set of
good quality tools. Quality is of prime importance since cheap
tools will prove expensive in the long run if they slip or break

when in use, causing personal injury or expensive damage to
the component being worked on. A good quality tool will last a
long time, and more than justify the cost.

For practically all tools, a tool factor is the best source since
he will have a very comprehensive range compared with the
average garage or accessory shop. Having said that, accessory
shops often offer excellent quality tools at discount prices, so it
pays to shop around. There are plenty of tools around at
reasonable prices, but always aim to purchase items which meet
the relevant national safety standards. If in doubt, seek the
advice of the shop proprietor or manager before making a
purchase.

The basis of any tool kit is a set of open-ended spanners,
which can be used on almost any part of the machine to which

there is reasonable access. A set of ring spanners makes a useful addition, since they can be used on nuts that are very tight or where access is restricted. Where the cost has to be kept within reasonable bounds, a compromise can be effected with a set of combination spanners – open-ended at one end and having a ring of the same size on the other end. Socket spanners may also be considered a good investment, a basic $3/8$ in or $1/2$ in drive kit comprising a ratchet handle and a small number of socket heads, if money is limited. Additional sockets can be purchased, as and when they are required. Provided they are slim in profile, sockets will reach nuts or bolts that are deeply recessed. When purchasing spanners of any kind, make sure the correct size standard is purchased. Almost all machines manufactured outside the UK and the USA have metric nuts and bolts, whilst those produced in Britain have BSF or BSW sizes. The standard used in USA is AF, which is also found on some of the later British machines. Others tools that should be included in the kit are a range of crosshead screwdrivers, a pair of pliers and a hammer.

When considering the purchase of tools, it should be remembered that by carrying out the work oneself, a large proportion of the normal repair cost, made up by labour charges, will be saved. The economy made on even a minor overhaul will go a long way towards the improvement of a toolkit. .

In addition to the basic tool kit, certain additional tools can prove invaluable when they are close to hand, to help speed up a multitude of repetitive jobs. For example, an impact screwdriver will ease the removal of screws that have been tightened by a similar tool, during assembly, without a risk of damaging the screw heads. And, of course, it can be used again to retighten the screws, to ensure an oil or airtight seal results. Circlip pliers have their uses too, since gear pinions, shafts and similar components are frequently retained by circlips that are not too easily displaced by a screwdriver. There are two types of circlip pliers, one for internal and one for external circlips. They may also have straight or right-angled jaws.

One of the most useful of all tools is the torque wrench, a form of spanner that can be adjusted to slip when a measured amount of force is applied to any bolt or nut. Torque wrench settings are given in almost every modern workshop or service manual, where the extent to which a complex component, such as a cylinder head, can be tightened without fear of distortion or leakage. The tightening of bearing caps is yet another example. Overtightening will stretch or even break bolts, necessitating extra work to extract the broken portions.

As may be expected, the more sophisticated the machine, the greater is the number of tools likely to be required if it is to be kept in first class condition by the home mechanic. Unfortunately there are certain jobs which cannot be accomplished successfully without the correct equipment and although there is invariably a specialist who will undertake the work for a fee, the home mechanic will have to dig more deeply in his pocket for the purchase of similar equipment if he does not wish to employ the services of others. Here a word of caution is necessary, since some of these jobs are best left to the expert. Although an electrical multimeter of the AVO type will prove helpful in tracing electrical faults, in inexperienced hands it may irrevocably damage some of the electrical components if a test current is passed through them in the wrong direction. This can apply to the synchronisation of twin or multiple carburettors too, where a certain amount of expertise is needed when setting them up with vacuum gauges. These are, however, exceptions. Some instruments, such as a strobe lamp, are virtually essential when checking the timing of a machine powered by CDI ignition system. In short, do not purchase any of these special items unless you have the experience to use them correctly.

Although this manual shows how components can be removed and replaced without the use of special service tools (unless absolutely essential), it is worthwhile giving consideration to the purchase of the more commonly used tools if the machine is regarded as a long term purchase Whilst the alternative methods suggested will remove and replace parts without risk of damage, the use of the special tools recommended and sold by the manufacturer will invariably save time.

Chapter 1 Engine

Contents

General description 1
Operations with engine in frame 2
Operations with engine removed 3
Method of engine removal 4
Dismantling the engine; removing the petrol tank 5
Removing the primary drive, clutch and chaincase 6
Removing the engine unit from the frame 7
Dismantling the engine: general 8
Dismantling the engine: removing the cylinder head and
cylinder barrel 9
Dismantling the engine: removing the piston 10
Dismantling the engine: removing the Magdyno or magneto
(all except alternator models) 11
Dismantling the engine: removing the contact breaker assembly
(alternator models only) 12
Dismantling the engine: removing the camshafts and drive
gears 13
Dismantling the engine: removing the camshaft spindles
and tappets 14
Dismantling the engine: removing the oil pump 15
Dismantling the engine: separating the crankcase halves ... 16
Examination and renovation: general 17
Main bearing and flywheel assembly: examination and
renovation 18
Small end bush: examination and renovation 19
Timing pinions: examination and renovation 20
Cylinder barrel: examination and renovation 21

Piston and rings: examination and renovation 22
Valves, valve springs and valve guides: examination and
renovation 23
Cylinder head: examination and renovation 24
Rockers, rocker spindles and rocker covers: examination
and renovation (all ohv models) 25
Engine reassembly: general 26
Engine reassembly: jointing the crankcases 27
Engine reassembly: replacing the oil pump and crankcase
sump plate 28
Engine reassembly: replacing the camshaft and timing
pinions 29
Engine reassembly: fitting the piston and cylinder barrel ... 30
Engine reassembly: replacing the Magdyno (or magneto)
and retiming the ignition 31
Engine reassembly: refitting the contact breaker assembly
and timing the ignition (alternator models only) 32
Engine reassembly: replacing the cylinder head 33
Engine reassembly: setting the valve clearances 34
Replacing the engine unit in the frame 35
Engine reassembly: replacing the primary chain case
inner, the primary drive and clutch 36
Engine reassembly: replacing the alternator 37
Engine reassembly: completion 38
Starting and running the rebuilt engine 39
Fault diagnosis: engine 40

Specifications

BSA model:	M20	M21
Engine - general		
Type	Air-cooled, single cylinder, side-valve four-stroke	
Cylinder barrel	Cast iron	
Bore	82 mm	82 mm
Stroke	94 mm	112 mm
Capacity	496 cc	591 cc
Bhp	13 @ 4,200 rpm	15 @ 4,000 rpm
Compression ratio	4.9 : 1	5 : 1
Cylinder head	Aluminium alloy	
Piston		
Type	Aluminium alloy	
Clearance at bottom of skirt	0.004 - 0.006 in (0.1 - 0.15 mm)	
Piston rings		
No. of	Two compression, one oil control	
End gap	0.008 - 0.010 in (0.20 - 0.25 mm)	
Side clearance	0.002 - 0.004 in (0.05 - 0.10 mm)	
Valves		
Seat angle	45°	

Valve timing

Inlet opens BTDC	... ...	25°
Inlet closes ABDC	... ...	65°
Exhaust opens BBDC	... ...	65°
Exhaust closes ATDC	... ...	25°

Valve clearances (cold)

| Inlet | ... ... | 0.010 in (0.25 mm) |
| Exhaust | ... ... | 0.012 in (0.30 mm) |

BSA model:	M33	B31	B33
Engine - general			
Type	Air-cooled, single cylinder, overhead-valve four-stroke		
Cylinder barrel	Cast iron	Cast iron	Cast iron
Bore	85 mm	71 mm	85 mm
Stroke	88 mm	88 mm	88 mm
Capacity	499 cc	348 cc	499 cc
Bhp	23 @ 5,500 rpm	17 @ 5,500 rpm	23 @ 5,500 rpm
Compression ratio	6.8 : 1	6.5 : 1	6.8 : 1
Cylinder head	Cast iron	Cast iron	Cast iron
Piston			
Type	Aluminium alloy	Aluminium alloy	Aluminium alloy
Clearance at bottom of skirt ...	0.0045 - 0.0065 in (0.11 - 0.16 mm)	0.0040 - 0.0055 in (0.10 - 0.14 mm)	0.0045 - 0.0065 in (0.11 - 0.16 mm) (split skirt piston 0.0006 - 0.003 in [0.015 - 0.080 mm])

Piston rings

No. of	Two compression, one oil control
End gap	0.008 - 0.012 in (0.20 - 0.30 mm)
Side clearance	0.002 - 0.004 in (0.05 - 0.10 mm)

Valve springs

| Free length - inner ... ... | 1 13/16 in (46 mm) |
| Free length - outer ... ... | 2 5/32 in (54.75 mm) |

Valves

| Seat angle ... ... | 45° |

Valve timing

Inlet opens BTDC	25°
Inlet closes ABDC	65°
Exhaust opens BBDC	65°
Exhaust closes ATDC	25°
Valve clearances (cold)	
Inlet	0.003 in (0.08 mm)
Exhaust	0.003 in (0.08 mm)

BSA model:	B32	B34
Engine - general		
Type	Air-cooled, single cylinder, overhead-valve four-stroke	
Cylinder barrel	Aluminium alloy	
Bore	71 mm	85 mm
Stroke	88 mm	88 mm
Capacity	348 cc	499 cc
Bhp	17 @ 5,500	23 @ 5,500
Compression ratio	6.5 : 1	6.8 : 1
Cylinder head	Aluminium alloy	
Piston		
Type	Aluminium alloy	
Clearance at bottom of skirt	0.0040 - 0.0055 in (0.10 - 0.14 mm)	0.0045 - 0.0065 in (0.11 - 0.16 mm)

Piston rings
No. of   ...   .   Two compression, one oil control
End gap ...   .   0.008 - 0.012 in (0.20 - 0.30 mm)
Side clearance   ...   .   0.002 - 0.004 in (0.05 - 0.10 mm)

Valve springs
Free length - inner   ...   .   1 13/16 in (46 mm)
Free length - outer   ...   .   2 5/32 in (54.75 mm)

Valves
Seat angle   ...   .   45°

Valve timing
Inlet opens BTDC   ...   .   25°
Inlet closes ABDC   ...   .   65°
Exhaust opens BBDC   ...   .   65°
Exhaust closes ATDC   ...   .   25°

Valve clearances (cold)
Inlet   ...   .   0.003 in (0.08 mm)
Exhaust   ...   .   0.003 in (0.08 mm)

BSA model: B32 Gold Star   B34 Gold Star

Engine - general
Type   ...   .   Air-cooled, single cylinder, overhead-valve, four-stroke

Cylinder barrel
Cylinder barrel   ...   .   Aluminium alloy   Aluminium alloy
Bore   ...   .   71 mm   85 mm
Stroke   ...   .   88 mm   88 mm
Capacity   ...   .   348 cc   499 cc
Bhp   ...   .   Variable   Variable
Compression ratio:
    Touring   ...   .   CB 7.25 : 1; DB 8.0 : 1   CB 7.25 : 1; DB 8.0 : 1
    Scrambles   ...   .   CB 9.0 : 1; DB 9.0 : 1   CB 9.0 : 1; DB 8.75 : 1
    Racing   ...   .   CB 8.0 : 1; DB 9.0 : 1   CB 8.0 : 1; DB 8.0 : 1

Cylinder head
Cylinder head ...   .   Aluminium alloy   Aluminium alloy

Piston
Type   ...   .   Aluminium alloy   Aluminium alloy
Clearance at bottom of skirt:
    Touring   ...   .   0.0025 - 0.0045 in (0.06 - 0.11 mm)
    Competition ...   .   0.004 - 0.006 in (0.10 - 0.15 mm)

Piston rings
No. of   ...   .   Two compression, one oil control (scraper)
End gap:
    Top ring   ...   .   0.012 in (0.30 mm)
    2nd and oil scraper ...   .   0.010 in (0.25 mm)
Side clearance   ...   .   0.0005 - 0.001 in (0.013 - 0.025 mm)

Valve springs
Free length - inner   ...   .   1.75 in (44.4 mm)
Free length - outer   ...   .   2.156 in (54.75 mm)

Valves
Seat angle   ...   .   45°

Valve timing at 0.018 in (0.46 mm) valve clearance

	Inlet opens BTDC	Inlet closed ABDC	Exhaust opens BBDC	Exhaust closes ATDC
B32 Gold Star				
Touring	43°	73°	64°	34°
Scrambles	60°	85°	84° 50'	60° 10'
Racing	60°	85°	84° 50'	60° 10'
Clubman	60°	85°	84° 50'	60° 10'
B34 Gold Star				
Touring	43°	73°	70°	45°
Scrambles	63°	72°	80°	55°

Racing	65°	85°	80°	55°
Clubman	65°	85°	80°	55°

If special engine shaft pinion 65 - 696 is used valve timing will be 10° earlier.

Valve clearances (cold)

All models except Touring CB and DB Gold Star

Inlet ... ... 0.006 in (0.15 mm)
Exhaust ... ... 0.006 in (0.15 mm)

Touring CB and DB Gold Star

Inlet ... ... 0.008 in (0.20 mm)
Exhaust ... ... 0.010 in (0.25 mm)

Cams

BSA Part Number

	Inlet	Exhaust
B32 Gold Star		
Touring	65 - 2448	65 - 2450
Scrambles	65 - 2444	65 - 1891
Racing	65 - 2444	65 - 1891
Clubman	65 - 2444	65 - 1891
B34 Gold Star	*Inlet*	*Exhaust*
Touring	65 - 2448	65 - 2450
Scrambles	65 - 2446	65 - 2446
Racing	65 - 2442	65 - 2446
Clubman	65 - 2442	65 - 2446

1 General description

The engines fitted to the BSA M series, B series and Gold Star models are fundamentally similar units differing to a greater or lesser degree, depending on the model. Although the engines share many points in common, and in some cases utilise identical components, the range may be separated into three distinct categories. The M20 and M21 models of 500 cc and 600 cc respectively are both fitted with side-valve engines, using cast iron cylinder barrels and aluminium alloy cylinder heads. The B31 and B33 models have overhead valve engines fitted with cast iron cylinder heads and cylinder barrels. The M33 engine is identical to that of the B33, the difference in the two models being in variations of the cycle parts and finish. The B32 and B34 competition models and also the 350 cc and 500 cc Gold Star range utilise all alloy engines of overhead-valve configuration.

All models share the crankcase and lower engine design. The two aluminium alloy crankcase halves separate vertically in a fore and aft plane and house a crankshaft assembly comprising two full flywheels and mainshafts running on roller and journal ball bearings. The mainshafts are located in the flywheels by an interference fit and are secured by a number of rivets. The connecting rod big-end bearing is of the caged double row roller type, supported on a tapered fit crankpin which is secured by thin nuts and locking plates. On all but the Gold Star models, the crankpin is secured in the drive-side flywheel by a Woodruff key.

Ignition on most engines, including many Gold Star machines, is provided by a conventional magneto incorporated into the Lucas 'Magdyno' unit. Many B32 and B34 and Gold Star models were also fitted with independent Lucas or BTH magnetos, the dynamo being dispensed with for competition events. Without exception, the ignition is controlled manually by a slack-wire advance handlebar lever.

Carburation on the M series and B31-34 models is catered for in general by the Amal Monobloc instrument. Some 1954 models, however, utilised the Standard Amal carburettor which had a separate float chamber. Gold Star machines were fitted with 10TT9, RN or GP Amal instruments, depending on the model and function of the machine when it left the factory. Some CB engined Gold Star scrambles machines were fitted with the type 376 or 389 Monobloc instruments.

Lubrication is effected on the dry sump principle in which oil is fed by gravity to a gear-type pump and distributed to various parts of the engine, after which it drains to the crankcase sump and is returned to the oil tank by the scavenge pump. Both the feed pump and scavenge pump are of the gear-type, contained in a common pump block and driven by the same shaft from a worm wheel integral with and inboard of the half-time pinion (engine shaft pinion). The lubrication system is protected by two filters and a pressure release ball-valve. The main filter, which is of the gauze or wire mesh type (depending on the model), is contained within the oil tank. The secondary filter, which is of the gauze type, is fitted above the crankcase sump plate and prevents larger particles of matter from passing through the return side of the pump.

2 Operations with engine in frame

1 It is not necessary to remove the engine from the frame unless the crankshaft assembly requires attention. Most operations can be accomplished with the engine in the frame, such as:

a) *Removal and replacement of the cylinder head.*
b) *Removal and replacement of the cylinder barrel and piston.*
c) *Removal and replacement of the clutch and primary drive.*
d) *Removal of the timing pinions.*
e) *Removal of the oil pump.*
f) *Removal of the dynamo and/or magneto.*

2 When several operations have to be undertaken simultaneously, such as during an extensive rebuild or overhaul, it is often advantageous to remove the engine from the frame after some preliminary dismantling. This will give the advantage of better access and more working space, especially if the engine is attached to a bench-mounted stand.

3 Operations with the engine removed

1 Removal and replacement of the main bearings.
2 Removal and replacement of the crankshaft assembly.

4 Method of engine removal

1   Removal of the engine from the frame is straightforward and since the gearbox is a completely separate unit may be accomplished without the need to disturb it. Removal of the gearbox may also take place without disturbing the engine.

2   The engine may be lifted from the frame as a unit without need to remove any of the more major components. It is, however, heavy, requiring two people to manoeuvre it from place and in consequence it may be considered easier to remove the cylinder barrel, cylinder head and piston first. This may be carried out with ease after removing the petrol tank, exhaust pipe and carburettor, and then following the procedure in Section 9.

5 Dismantling the engine: removing the petrol tank

1   Place the machine on the centre stand and ensure that it is standing firmly on level ground.

2   Turn off the petrol supply and disconnect the petrol pipe by unscrewing the union where it joins the base of the petrol tap.

3   Remove the rubber plug from the centre of the tank top and use a box or socket spanner to slacken and remove the retaining bolt. Remove also the metal strip that braces both halves of the petrol tank. This is located on the underside of the tank, close to the steering head. It is secured by two small nuts. The tank can now be lifted clear from the frame.

The tank fitted to the M series machines is retained in a different manner. The front of the tank is secured by two bolts, one of which passes into each side of the steering head lug through projections from the nose of the tank. The rear of the tank is retained by bolts passing upwards into the tank base, through lugs attached to the top tube of the frame. It is necessary to remove also the bolt through the nose of the saddle, so that the saddle can be raised to allow sufficient clearance for removal of the petrol tank.

6 Dismantling the engine: removing the primary drive, clutch and chaincase

1   Place a shallow drain pan under the primary chaincase and remove the drain plug. On models fitted with a pressed steel chaincase, the drain plug is situated to the rear of and below the

footrest. The drain plug on other models is incorporated in one of the lower casing screws. It is the second screw from the front, in the bottom run.

2   Remove the left-hand footrest after unscrewing the single retaining nut. It may be necessary to ease the footrest off the shaft with a rawhide mallet. Loosen the footbrake pedal pinch bolt and draw the pedal off the splined shaft.

3   Loosen the fifteen chaincase screws (16 on M series) in an even and diagonal sequence. Pull the casing away from the inner case, allowing the remainder of the lubricant to flow into the drain pan. The steel chaincase is sealed by a thick rubber or cork joint. This should be preserved if in good condition as it may be used again. The gasket used on aluminium cases should be discarded and a new one fitted on reassembly.

4   On alternator B31 and B33 models the alternator rotor and stator may now be removed. Trace the main lead from the alternator upwards until access may be gained to the three snap connectors. Disconnect the wires from the connectors so that they may be pulled through the grommet in the centre of the inner chaincase. This task is eased if each separate wire is pulled through independently.

5   The six-coil stator is supported on three studs projecting from the inner chaincase, and is retained by three nuts and spring washers. After removing the nuts carefully ease the stator from position, taking care not to snag the main electrical lead. Note and remove the distance piece fitted on each stud. Knock the tab washer down, away from the rotor retaining nut, and loosen the nut. To prevent rotation of the crankshaft while the nut is loosened, place the machine in top gear and apply the rear brake. After removal of the nut and washer, the rotor may be pulled off the shaft end. Remove the Woodruff key from the keyway in the shaft.

6   On all but the alternator models a spring loaded cam type engine shock absorber is fitted to the left-hand mainshaft. Commence dismantling the shock absorber by loosening the special ring nut. A tab washer is fitted between the ring nut and the splined shock absorber sleeve. Bend the tab washer away from the ring nut where it was bent inwards on assembly to secure the nut. The ring nut may be very tight and consequently loosening by means of a 'C' spanner may not be possible. A suitable steel drift should be used to loosen the nut after preventing crankshaft rotation by placing the machine in top gear and applying the rear brake.

After nut removal the component parts of the shock absorber may be withdrawn.

7   Loosen the clutch spring locknuts followed by the adjuster

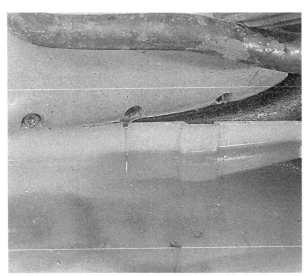

6.1a Drain primary chaincase oil followed by ...

6.1b ... the oil tank and ...

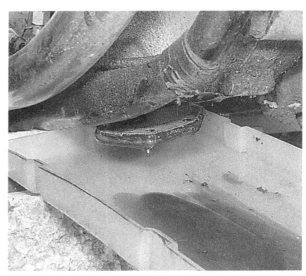

6.1c ... the crankcase

6.1d Remove the sump plate and filter screen

6.3 Detach the primary chaincase cover

6.6 Loosen and unscrew the shock absorber nut and then ...

6.7 ... remove the clutch spring adjuster nuts

6.8a Remove the clutch pressure plate

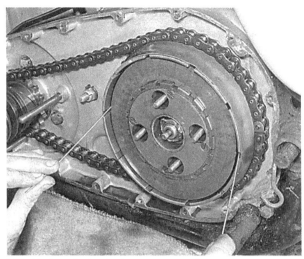

6.8b Hook each clutch plate out separately

6.9a Loosen and remove clutch centre nut with a box spanner

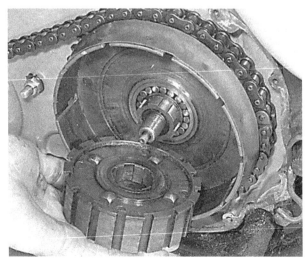

6.9b Pull clutch centre boss from shaft

nuts, springs and spring cups. On alternator models only four springs are fitted, and the place of the nuts is taken by four slot headed sleeve nuts. To aid removal of the sleeve nuts, a special slotted screwdriver may be fabricated from mild steel plate, suitably shaped, so that the studs do not foul the screwdriver blade. This tool, once fabricated, is invaluable as it can be used for nut removal and clutch spring adjustment.

8 Remove the clutch pressure plate followed by the plain and friction clutch plates, noting their order. If difficulty is encountered in their removal, use two wire hooks to draw them from the clutch outer drum.

9 Prevent the clutch shaft from rotating by engaging top gear and applying the rear brake. Knock down the tab washer and remove the clutch centre nut. A socket or box spanner should be used. These nuts are often mistreated by the application of unsuitable spanners in tightening or loosening. If the nut is so damaged that a spanner will not fit correctly, a punch or blunt edged cold chisel may be used for loosening. This should only be attempted on a nut already damaged and therefore in need of renewal. Pull the clutch centre off the shaft.

10 On all models the clutch outer drum and the engine sprocket may be removed simultaneously, whilst still in mesh with the primary drive chain. It is also possible to separate the primary drive chain by means of the spring link and then remove the two sprockets independently. Before separating the chain, rotate the engine and clutch outer drum so that the spring link is on the lower run adjacent to the recess in the rear of the casing. This will enable the master link to be removed without fouling the inner casing.

11 When removing the clutch outer drum from the alternator type engine, note that the clutch bearing is of the uncaged roller type and that as the drum is removed the twenty (20) rollers will fall free. Make arrangements to catch the rollers in a suitable container placed below the chaincase. On other models the clutch bearing consists of two separate caged ball races, supported on a double row inner race. These may be removed from the clutch shaft followed by the clutch drum backing plate.

12 The splined sleeve upon which the clutch is mounted is a taper fit on the shaft and is located by a Woodruff key, consequently the sleeve will require drawing from position. BSA service tool No. 61-3362 should be used if it is available. If the correct tool is not to hand, a two or three legged sprocket puller may be used instead. The legs of the sprocket puller should be located to the rear of the boss on the clutch sleeve. It may be necessary to remove the three wired-up bolts passing through the chaincase inner into the crankcase, and on aluminium alloy cases, the single external casing bolt to the rear, to give enough clearance to fit the sprocket puller. Before fitting the sprocket puller, replace the shaft nut to prevent the threaded portion of the shaft from spreading under load.

The clutch sleeve is often extremely tight and will therefore resist removal. If this is the case, tighten the puller down firmly and then give the puller bolt a single heavy blow with a hammer. This invariably frees the mating components. After removal of the clutch sleeve, prise the Woodruff key from the shaft.

13 The inner chaincase can now be removed, after unwiring and removing the three forward bolts and on aluminium alloy cases removing the external bolt which passes through a lug on the rear of the case and into the frame. Note the steel plate fitted to the crankcase boss on some models.

7 Dismantling the engine: removing the engine unit from the frame

1 Place a container of 1 pint capacity below the crankcase and remove the drain plug. Allow the oil to drain and then remove the sump plate, which is retained by four nuts and washers. The secondary oil filter (gauze screen) will come away with the sump plate. Position a larger container below the oil tank drain plug. On M series machines the plug is in the base of the oil tank. On other models the drain plug screws into the lower portion of the

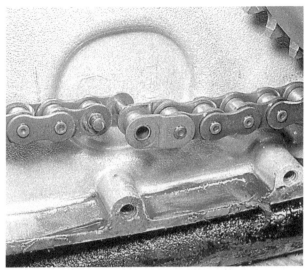

6.10a Disconnect the primary chain at spring link

6.10b Remove shock absorber as a unit

6.10c Note spacer behind shock absorber unit

6.11 Lift clutch outer drum from shaft bearing

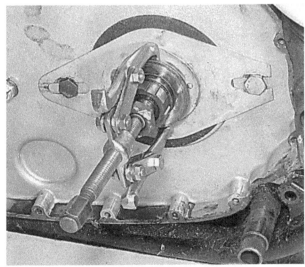

6.12 The clutch sleeve must be drawn from position

6.13a Primary drive inner case held by three forward bolts and ...

6.13b ... single bolt passing through frame lug at the rear

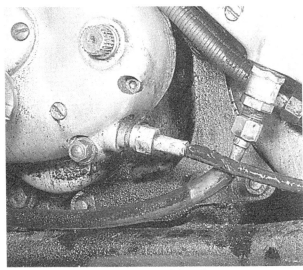

7.2a Disconnect the speedometer cable and ...

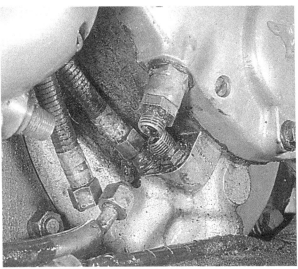

7.2b ... oil feed and return pipes and breather pipe

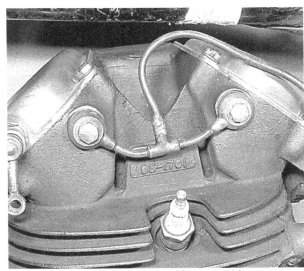

7.3 Detach the rocker oil feed pipes (OHV models)

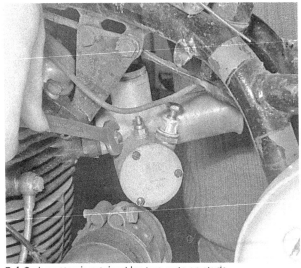

7.4 Carburettor is retained by two nuts on studs

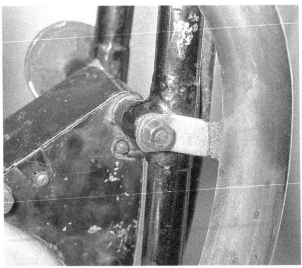

7.5a Exhaust system held by front engine mounting bolt and ...

tank, and incorporates a cylindrical wire mesh filter. When draining either machine, it is worthwhile arranging a cardboard or tin
shute down which the oil may flow into the drain pan. Up to 5
pints may be released, so use a container of large enough
capacity.

2   Unscrew the timing chest breather pipe from the lower rear
edge of the casing. On all M and B series machines the pipe is of
the olive type, retained by a packing nut. On Gold Star and
competition machinery a banjo union is used. Detach the speedometer cable at the gearbox by unscrewing the cable retaining nut.
Disconnect the oil feed pipe and oil return pipe from the adaptors
in the right-hand crankcase half.

3   Loosen and remove the two banjo bolts which connect the
rocker oil feed to the left-hand end of the rocker shafts (ohv
models only).

4   Unscrew the carburettor cap and pull the complete throttle
valve/choke slide assembly from the carburettor. Tie the
assembly and cables to some part of the frame so that they are
out of the way and will not become damaged during subsequent
dismantling operations. On competition carburettors (TT and GP
types) the choke assembly consists of a cable operated plunger
remote from the main throttle valve and carburettor cap.
Unscrew the housing nut to release the cable and plunger.
    Slacken the two carburettor retaining nuts and then remove
them. On competition carburettors fitted with remote float
chambers disconnect the chamber from the rubber mounting.
The air filter (if fitted) may require removal before the carburettor is pulled off the studs.

5   Remove the silencer retaining bolt and the exhaust pipe
retaining bolt (where fitted). The silencer is often secured by the
right-hand pillion footrest. Carefully ease the complete exhaust
system from position in the exhaust port. Take especial care not
to allow the pipe end to tie in the exhaust port as damage to the
cylinder head may result. If the pipe is very tight due to a build
up of carbon, use a rawhide mallet to tap the inner curve of the
pipe away from the cylinder head.

6   On alternator models detach the suppressor cap from the
spark plug and secure the cap and HT lead out of harms way.
Disconnect the low tension lead at the contact breaker housing
by removing the brass nut on the terminal. On all other models
disconnect the two leads at the dynamo. On early dynamos, the
two leads are retained by a backing plate in a socket incorporated
in the end cover. On later models remove the end cover first,
which is held by a single screw to gain access to the wires which
are retained in a manner similar to that of the early dynamos.

7   Remove the engine steady bracket, which is retained by a
single bolt to a cast-in lug on the rear of the cylinder head and

7.5b ... by a single bolt at rear (in this case the pillion footrest)

7.5c Separate pipe from silencer, if necessary

7.6 Disconnect the dynamo leads (Magdyno models)

7.7a Remove the two rear engine mounting bolts and ...

by one or two bolts (depending on the model) to a leg welded to
the frame. Loosen and remove the nuts and washers from the
engine mounting studs on one side only. Remove all the mounting studs except the lower front stud, which passes through the
two tubular lugs welded to the frame, and the stud directly
below the crankcase which cannot be removed until the engine
has been raised slightly. Remove (where fitted) the front engine
mounting plate cover. On some models the cover is retained by
the upper mounting studs, and on other models by four pan-
head screws. When detaching the engine plate cover, note and
remove the various spacers that are fitted to the studs. These
will in any case probably fall out when the studs are removed.
8   Disconnect the exhaust lifter cable at the operating arm by
applying a suitable lever to the arm, or by screwing the cable
adjuster inwards to gain the necessary slack in the cable. Where
fitted, unscrew the rev-counter drive cable at the timing chest.
9   Before lifting the engine from the frame, ensure that no
bolts or ancillary components are still in place which may prevent
easy removal. Check that any cables and wires are tucked out of
harms way. The engine is a heavy unit and, by virtue of its shape,
is rather difficult to lift. It is advised that two people perform
the final removal operation to prevent damage to the machine or
to the operator. Lift the engine upwards at the front and push
out the main mounting stud. Raise the engine further and remove
the final stud, previously obscured by the frame tubes. The
engine may now be tilted to either the right or the left and
lifted clear of the frame.

8   Dismantling the engine: general

1   Before commencing work on the engine unit, the external
surfaces should be cleaned thoroughly. A motorcycle engine has
very little protection from road grit and other foreign matter,
which will find its way into the dismantled engine if this simple
precaution is not observed. One of the proprietary cleaning
compounds such as Gunk can be used to good effect, particularly if the compound is allowed to work into the film of oil
and grease before it is washed away. When washing down, make
sure that water cannot enter the carburettor or the electrical
system, particularly if these parts have been exposed.
2   Never use undue force to remove any stubborn part, unless
mention is made of this requirement. There is invariably good
reason why a part is difficult to remove, often because the
dismantling operation has been tackled in the wrong sequence.
Dismantling will be made easier if a simple engine stand is

constructed that will correspond with the engine mounting
points. This arrangement will permit the complete unit to be
clamped rigidly to the workbench, leaving both hands free.

9   Dismantling the engine: removing the cylinder head and cylinder barrel

Side-valve models only

1   The cylinder head fitted to the side-valve engines is of cast
aluminium alloy and is retained by ten bolts and plain washers.
Before loosening the head bolts, slacken off the spark plug.
Loosen the cylinder head bolts in as much of a diagonal
pattern as possible, to prevent warpage. The cylinder head
can be lifted from position after removal of all the bolts. If
the head is firmly stuck to the gasket, tap the underside of the
head with a rawhide mallet. Take care not to damage the fins,
which may fracture if overstressed.
2   Rotate the crankshaft so that the piston is at TDC on
the compression stroke. Loosen and remove the five cylinder
barrel retaining nuts and washers. Using a rawhide mallet, tap
the cylinder barrel upwards, to release the base flange from the
base gasket. Only apply the mallet to the areas below the inlet
port and exhaust port which, unlike the fins, are less likely to
suffer damage. When the cylinder barrel is free, lift it upwards
off the studs and the piston. If a top-end overhaul only is
contemplated, place a clean rag around the connecting rod in
the crankcase mouth before the piston rings leave the cylinder
bore. This will prevent particles of broken piston ring from
falling into the crankcase and will subsequently prevent foreign
matter from finding its way into the engine.
     Support the piston and connecting rod as the piston leaves
the cylinder or it will drop under its own weight and may
strike a barrel stud or the edge of the crankcase.

M33, B31 and B33 ohv models only

1   Detach the oil return pipe which leads from the cylinder
head to the crankcase, by removing the two banjo bolts.
Slacken the castellated gland nut which holds the pushrod
cover tower to the cylinder head. In an emergency, a soft
brass drift may be used to loosen the gland nut but great care
must be taken not to bruise the soft aluminium alloy. The
correct tool for loosening the gland nut is a close fitting 'C'
spanner, which was supplied in the original tool kit. It is
well worth acquiring a suitable spanner or fabricating one

7.7b ... all the bolts retaining the front plates

9.1 Detach the rocker oil return pipe (B Series iron head)

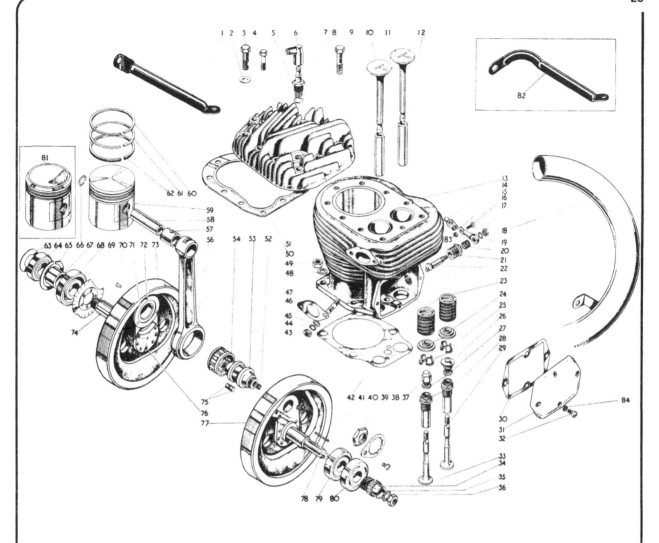

Fig. 1.1. Cylinder head, barrel and flywheel assembly - side-valve models - component parts

1	Engine stay	23	Valve spring - 2 off	45	Washer - 2 off	67	Distance piece
2	Washer - 9 off	24	Valve spring collar - 2 off	46	Hallite gasket	68	Drive side roller
3	Bolt - 5 off	25	Valve spring collet - 2 off	47	Stud - 2 off	69	Oil thrower
4	Bolt	26	Tappet head	48	Spring washer - 5 off	70	Woodruff key
5	Spark plug	27	Tappet locknut - 2 off	49	Nut - 5 off	71	Thrust washer
6	Spark plug cap	28	Tappet guide - 2 off	50	Cylinder head gasket	72	Rivet - 6 off
7	Spark plug washer	29	Exhaust tappet	51	Cylinder head	73	Drive side flywheel
8	Bolt - 4 off	30	Gasket	52	Screwed plug	74	Drive side mainshaft
9	Inlet valve guide	31	Inspection cover	53	Crankpin	75	Big end roller - 24 off
10	Inlet valve	32	Screw - 4 off	54	Bearing cage	76	Rivet - 14 off
11	Exhaust valve	33	Inlet tappet	56	Connecting rod	77	Timing side flywheel
12	Exhaust valve guide	34	Half time pinion	57	Small end bush	78	Timing side mainshaft
13	Barrel	35	Locking washer	58	Gudgeon pin	79	Inner roller bearing
14	Cable stop	36	Nut	59	Piston	80	Outer ball bearing
15	Exhaust lifter lever	37	Inlet tappet head	60	1st compression ring	81	Piston assembly
16	Screw	38	Screw	61	2nd compression ring	82	Alternative engine stay
17	Nut	39	Locking plate	62	Scraper ring		(RH sidecar only)
18	Exhaust pipe	40	Nut - 2 off	63	Shim	83	Screw
19	Washer	41	Woodruff key	64	Drive side roller bearing	84	Spring washer - 4 off
20	Spring	42	Cylinder base gasket	65	Spring ring		
21	Exhaust lifter body	43	Nut - 2 off	66	Circlip		
22	Exhaust lifter spindle	44	Washer - 2 off				

Engine

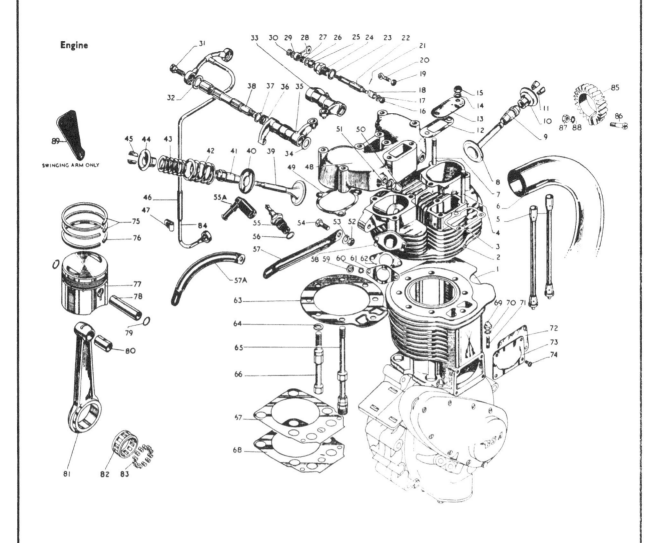

Fig. 1.2. Cylinder head and barrel - B32 and B34 Competition models - component parts

1	Barrel	24	Bush	47	Oil pipe clip	68	Cylinder base basket
2	Cylinder head	25	Felt washer	48	Rocker box	69	Domed nut - 2 off
3	Gasket	26	Washer	49	Gasket - 2 off	70	Washer - 2 off
4	Stud - 3 off	27	Spring	50	Nut - 2 off	71	Stud - 2 off
5	Pushrod - 2 off	28	Exhaust lifter lever	51	Washer - 2 off	72	Gasket
6	Exhaust pipe	29	Washer	52	Nut	73	Inspection cover
7	Stud - 6 off	30	Nut	53	Washer	74	Screw - 4 off
8	Exhaust valve	31	Banjo bolt - 2 off	54	Bolt	75	Compression rings - 2 off
9	Valve guide	32	Fibre washer	55	Spark plug	76	Scraper ring
10	Collar - 2 off	33	Valve rocker	55a	Spark plug cap	77	Piston
11	Collar - 2 off	34	Thrust washer - 2 off	56	Spark plug washer	78	Gudgeon pin
12	Gasket	35	Valve rocker	57	Engine stay	79	Circlip - 2 off
13	Rocker box inspection cover	36	Spring - 2 off	58	Stud - 2 off	80	Small end bush
14	Washer - 9 off	37	Washer - 2 off	59	Nut - 2 off	81	Connecting rod
15	Nut - 9 off	38	Rocker arm - 2 off	60	Washer - 2 off	82	Bearing cage .
16	Nut	39	Inlet valve	61	Tufnol gasket	83	Bearing rollers - 24 off
17	Washer	40	Valve spring seat	62	Paper gasket - 2 off	84	Rocker oil feed pipe
18	Exhaust lifter	41	Valve guide	63	Cylinder head gasket	85	Exhaust pipe clamp
19	Nut	42	Valve spring outer - 2 off	64	Washer - 4 off	86	Screw
20	Cable stop	43	Valve spring inner - 2 off	65	Cylinder base bolt - 4 off	87	Nut
21	Split pin	44	Collar	66	Cylinder and head bolt - 4 off	88	Washer
22	Spindle	45	Cotter - 2 off	67	Shim	89	Engine steady plate
23	Fibre washer	46	Rubber oil hose				

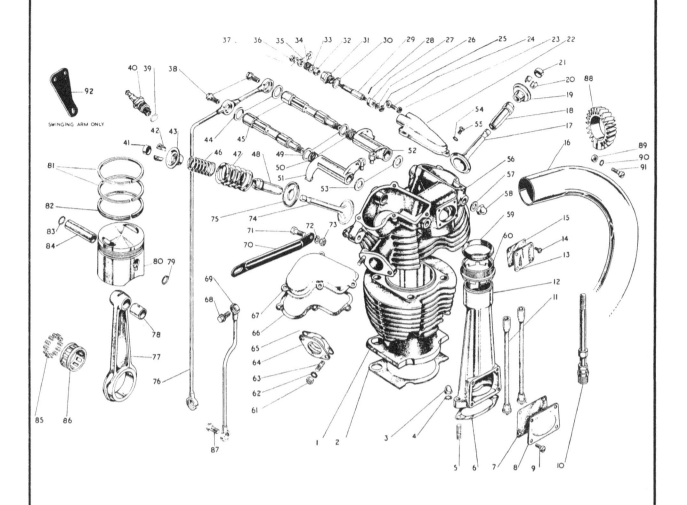

Fig. 1.3. Cylinder head and barrel - component parts

1 Barrel	24 Cable stop	47 Outer valve spring - 2 off
2 Shim	25 Nut	48 Valve guide
3 Domed nut - 2 off	26 Washer	49 Washer - 2 off
4 Washer - 2 off	27 Exhaust valve lifter	50 Spring - 2 off
5 Stud - 2 off	28 Split pin	51 Rocker, inlet
6 Gasket	29 Spindle	52 Rocker, exhaust
7 Gasket	30 Fibre washer	53 Thrust washer - 2 off
8 Inspection cover	31 Bush	54 Washer - 8 off
9 Screw - 4 off	32 Felt washer	55 Screw - 8 off
10 Cylinder base bolt - 4 off	33 Washer	56 Cylinder head
11 Pushrod - 2 off	34 Spring	57 Washer - 2 off
12 Pushrod tube	35 Exhaust lifter lever	58 Domed nut - 2 off
13 Inspection cover	36 Nut	59 Pushrod tube seal
14 Screw - 4 off	37 Washer	60 Pushrod tube nut
15 Gasket	38 Banjo bolt - 2 off	61 Nut - 2 off
16 Exhaust pipe	39 Spark plug washer	62 Washer - 2 off
17 Exhaust valve	40 Spark plug	63 Stud
18 Valve guide	41 End cap	64 Tufnol gasket
19 Collar	42 Cotter - 2 off	65 Paper gasket - 2 off
20 Cotter - 2 off	43 Collar	66 Rocker box cover gasket - 2 off
21 End cap - 4 off	44 Fibre washer - 2 off	67 Rocker box cover
22 Rocker box cover	45 Rocker spindle - 2 off	68 Banjo bolt
23 Nut	46 Inner valve spring - 2 off	69 Oil return pipe

70 Engine stay
71 Bolt
72 Washer
73 Nut
74 Inlet valve
75 Inlet valve spring seat - 2 off
76 Rocker oil feed pipe and unions
77 Connecting rod
78 Small end bush
79 Circlip
80 Piston
81 Compression ring - 2 off
82 Scraper ring
83 Circlip
84 Gudgeon pin
85 Crankpin rollers - 24 off
86 Bearing cage
87 Banjo bolt
88 Exhaust pipe clamp
89 Nut
90 Washer
91 Screw
92 Engine steady plate

9.3a Unscrew the cylinder head bolts

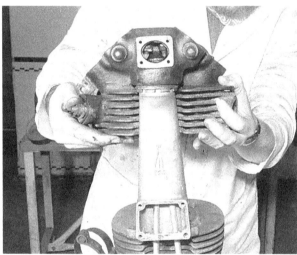
9.3b Lift cylinder head from position and ...

from mild steel plate, to prevent damage to the component during successive dismantling and reassembly operations.

2   Remove the tappet inspection cover from the bottom of the pushrod tower and the rocker access plate from the cylinder head. Both are retained by four screws. Loosen and remove the two acorn nuts and star washers from the pushrod tower bottom flange.

3   Slacken the spark plug to make easier its subsequent removal with the cylinder head detached. The cylinder head and cylinder barrel are retained by four long bolts, which are retained in the crankcase by bolt sockets and screw directly into the cylinder head material. Slacken and unscrew the four bolts by applying a spanner to the small hexagon on each shank. If the engine has not been dismantled recently, the bolts may be very tight, and owing to the smallness of the hexagons, little purchase can be gained. In addition to this, only an open-ended spanner can be used. The application of two spanners to the hexagon may aid loosening. Remember that the bolts should be unscrewed in a clockwise direction, when viewed from above. If this method fails, the sockets which retain each bolt may be loosened slightly to release the tension and then the bolts can be unscrewed in the normal manner. Do not loosen the sockets unless absolutely necessary as they will then require complete removal and reassembling with sealing compound to prevent oil leakage when the engine is in service. In extreme cases the sockets may resist slackening using the normal procedure. If great care is exercised, the area immediately adjacent to the sockets may be heated, using a blowtorch. This will expand the aluminium crankcase locally and release the sockets. This is a delicate operation; too little heat will have no effect; but too much may damage the crankcase.

4   After removal of the four long bolts the cylinder head may be lifted from position, complete with the pushrod tower. Prepare to catch the pushrods as the open lower end of the tower leaves the tappets. A rawhide mallet may be used to separate the cylinder head from the upper face of the cylinder barrel, if a build-up of carbon deposits causes sticking. As can be seen, no cylinder head gasket is fitted to these models, the gastight seal being effected by a ground joint.

5   The gland nut can now be unscrewed fully and the pushrod tower and sealing ring detached from the cylinder head.

6   Using a rawhide mallet, break the seal between the cylinder barrel and the crankcase. Take care to strike the barrel only at a point where the fins are strengthened by means of vertical webs or they may become damaged.

7   Lift the cylinder barrel upwards along the holding down

9.5 ... then unscrew pushrod tunnel gland nut

9.7 Carefully ease the cylinder barrel off the piston

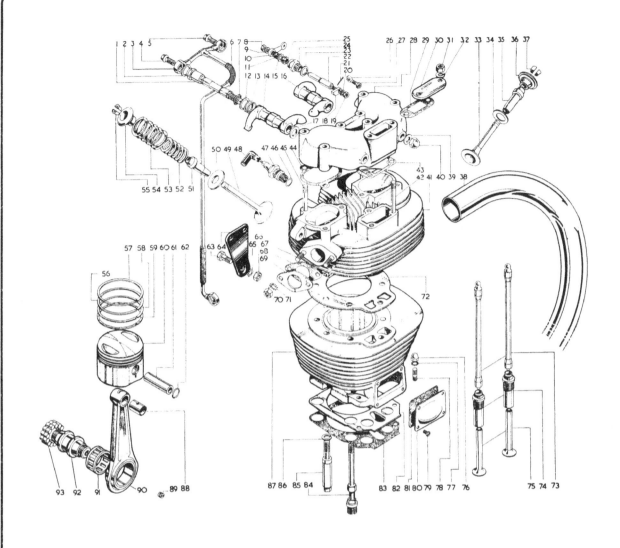

Fig. 1.4. Cylinder head and barrel - Gold Star models - component parts

1	Rocker arm spindle - 2 off	25	Felt washer
2	Fibre washer - 2 off	26	Cable stop
3	Rocker oil feed pipe	27	Nut
4	Inlet rocker feed bolt	28	Rocker box
5	Exhaust rocker feed bolt	29	Gasket
6	Oil feed hose - 2 off	30	Cover plate
7	'O' ring - 2 off	31	Nut - 9 off
8	Nut	32	Spring washer - 9 off
9	Washer	33	Exhaust valve
10	Exhaust lifter lever	34	Washer
11	Spring	35	Valve guide
12	Washer - 2 off	36	Spring collar
13	Spring - 2 off	37	Split cotter
14	Inlet rocker arm	38	Exhaust pipe
15	Washer	39	Washer - 2 off
16	Exhaust rocker arm	40	Domed nut - 2 off
17	Thrust washer - 2 off	41	Cylinder head
18	Exhaust lifter	42	Stud - 3 off
19	Nut	43	Gasket
20	Washer	44	Gasket - 2 off
21	Split pin	45	Stud - 6 off
22	Exhaust lifter spindle	46	Washer
23	Fibre washer	47	Spark plug
24	Bush	48	Plug cap

49	Inlet valve	73	Pushrod - 2 off
50	Washer - 2 off	74	Tappet guide - 2 off
51	Valve guide	75	Tappet - 2 off
52	Inner valve spring - 2 off	76	Domed nut - 2 off
53	Outer valve spring - 2 off	77	Washer - 2 off
54	Spring collar	78	Stud - 2 off
55	Split cotter - 2 off	79	Screw - 4 off
56	Bottom compression ring	80	Inspection plate
57	Top compression ring	81	Gasket
58	Bottom scraper ring	82	Shim
59	Top scraper ring	83	Gasket
60	Piston	84	Cylinder base fixing bolt - 4 off
61	Gudgeon pin	85	Cylinder head bolt - 4 off
62	Circlip - 2 off	86	Spring washer - 4 off
63	Bolt	87	Barrel
64	Engine steady plate	88	Small end bush
65	Washer	89	Grub screw
66	Nut	90	Connecting rod
67	Carburettor stud	91	Bearing cage
68	Paper gasket	92	Crankpin
69	Tufnol gasket	93	Crankpin rollers - 24 off
70	Nut - 2 off		
71	Washer - 2 off		
72	Cylinder head gasket		

10.1 Press out gudgeon pin to release the piston

11.2a Remove timing chest cover and old gasket

11.2b Note oil feed nozzle which is damaged easily

11.5 Remove magneto pinion nut

11.6 Drift Magdyno from pinion (emergency procedure)

11.7a Remove magneto pinion; note worn oil seal

bolts so that the piston begins to leave the cylinder bore. If a top-end overhaul only is anticipated, place a clean rag in the crankcase mouth around the connecting rod, before the piston rings leave the confines of the barrel. This will prevent pieces of broken piston ring from falling into the crankcase. The rag will also prevent the ingress of foreign matter during subsequent work. Support the piston and connecting rod as the piston leaves the barrel to prevent the piston falling under its own weight and sustaining damage.

B32, B34 and Gold Star models
1   The general procedure for cylinder head and barrel removal is fundamentally the same as that for the M and B series overhead valve engines. On competition and Gold Star engines the cylinder head is retained by eight bolts. On all but the CB type engines four long bolts are fitted, similar to those used on the cast-iron engines, and four short bolts, which pass up through the fins of the barrel. On most CB, DB and DBD engines there are three short bolts and five long bolts, one of which passes down the centre of the pushrod tunnel.
2   A one piece detachable rocker cover is fitted to all the Gold Star and competition cylinder heads. This does not require removal when detaching the cylinder head. The pushrods may be removed after the cylinder head has been lifted slightly, through the orifice covered by an access plate which is retained by two of the rocker cover retaining nuts.
3   Two specially shaped spanners are required for removing the cylinder head and barrel. The first is a heavily cranked ring spanner with which to remove the rear short cylinder head bolt. The outer periphery of the ring will have to be ground down so that it is thin enough to fit. The second spanner was originally included in the tool kit and consists of an open-ended ¼ in Whitworth spanner so bent near the head that it would reach the right-hand rear long bolt. In both cases suitable spanners could be provided by modifying existing tools. The open-ended spanner would require heating to a dull red, using a gas torch, before it can be bent.
4   If difficulty is experienced with tight bolts the same procedure as that used for the M and B series engines may be adopted. Bear in mind that as all the major components are manufactured in aluminium alloy, less force may be used and greater care must be exercised when applying external heat.
5   After cylinder head removal the rocker box should be removed. The box is held in place by nuts on studs.

10 Dismantling the engine: removing the piston

1   Remove one of the circlips from the piston and drive the gudgeon pin out of position whilst the piston and connecting rod is supported. If the gudgeon pin is a tight fit, warm the piston by placing a rag soaked in warm water on the crown. This should expand the gudgeon pin bosses sufficiently to release their grip on the gudgeon pin.
     It is important when drifting the gudgeon pin out that the connecting rod is adequately supported from the imposed side thrust. This can be done with a suitable block of wood. If this precaution is not taken, the connecting rod, or the bearing upon which it moves may be damaged.
2   When the piston is free, remove the other circlip and discard both. They should never be reused. Mark the piston on the inside of the skirt so that it is replaced in the same position. This is especially important if the crown has valve cutaways of equal depth and there is no obvious means of determining which is the front and rear.

11 Dismantling the engine: removing the Magdyno or magneto. All except alternator models

1   Most machines are fitted with a Lucas 'Magdyno', an instrument which combines a magneto to provide the ignition spark and a 6 volt DC generator. Machines intended purely for competition are fitted with a magneto only. In either case the

procedure for removal is identical.
2   Slacken the nine timing chest cover screws in an even and diagonal sequence. Remove the screws and lift off the cover. If the cover is firmly in place on the gasket, use a rawhide mallet to release the joint. On post-1955 Gold Star engines a small oil feed nozzle is fitted to the inside of the timing chest cover. To prevent damage to the nozzle, which feeds the main-shaft, leave three or four screws, equally spaced around the casing and partially unscrewed so that the joint seal is broken in a controlled manner and the nozzle is therefore not jarred or struck.
3   On the Gold Star engines, a timed breather is fitted to the outer face of the magneto drive pinion, from which it is driven by a pin offset on the face. The breather is spring loaded and will come away as the timing chest cover is removed.
4   On all models, an oil breather pipe is connected to the lower edge of the timing chest cover. If the breather is such that it will impede cover removal, detach the breather pipe from the union. The pipe is retained by a banjo bolt on a gland nut, depending on the model.
5   Apply a spanner to the magneto drive pinion centre nut. Jar the spanner in an anti-clockwise direction to loosen the nut. The drive pinion fits onto the tapered portion of the shaft extending from the Magdyno armature and will consequently require pulling from position using an extractor. The correct puller originally supplied for the job is BSA tool no. 61-1903. Removal of the Magdyno pinion is very difficult if the correct puller is not used. Because of the close proximity of the casing to the pinion, a standard sprocket puller cannot be utilised. The original BSA tool is still available from some motorcycle tool retailers, and it is advised that one is acquired. The tool screws into the centre of the pinion, pushing the pinion from the shaft when the centre screw is tightened.
6   An alternative method of pinion removal is possible, but this should be used only in an emergency. Loosen the strap which retains the Magdyno or magneto in position and pull the complete unit across so that the back face of the drive pinion is resting firmly against the inside face of the timing chest. Remove the pinion retaining nut and washer. Place a stout brass drift against the end of the Magdyno shaft and tap it smartly with a hammer to drive the shaft from the pinion. Considerable care should be taken to prevent damage to the shaft thread. This method is advised **only as a last resort** as in addition to danger of damage to the shaft end, the armature itself or the timing chest casting may be damaged.
7   After slackening the instrument retaining strap the unit

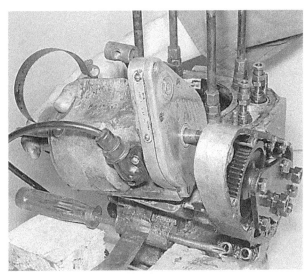
11.7b Lift Magdyno from position

13.2 Remove six bolts and detach steady plate

13.3 Check identification of timing marks before pinion removal

13.4a Half-time pinion must be drawn from shaft

may be lifted away complete. The strap is tensioned by a bolt passing through a trunnion captured in one part of the strap, into a similar threaded trunnion in the outer part of the strap. On some machines one or more shims may be found placed between the magneto base and the platform upon which the magneto rests. The shims are provided as an adjustment for meshing the magneto drive pinion and the idler pinion, and should not be lost.

12 Dismantling the engine: removing the contact breaker assembly. Alternator models only

1   Alternator models are fitted with a contact breaker assembly in place of the Magdyno, the operating cam of which is driven by a pinion similar to that used for driving the Magdyno of the earlier models. The assembly consists of a cast aluminium housing retained on the rear of the timing chest by the top three timing chest cover screws, which are secured by one nut each. The contact breaker unit is fitted into the housing and is retained by a clamping bracket, secured to the housing by a single bolt passing through a slightly elongated hole.
2   The complete assembly, including the drive pinion, may be detached from the rear of the casing after removal of the three cover screws and the nuts.
3   The drive pinion may now be detached, if required. The pinion is retained by a drive pin passing through a boss projecting from the front of the pinion, and through the driveshaft. The pin is prevented from movement by a spring clip, which locates in an annular groove in the boss. Prize the clip from position and push the pin from place, using a suitable tool.

13 Dismantling the engine: removing the camshafts and drive gears

1   On alternator models remove the timing chest cover as described in Section 11.2, if this has not already been done.
2   Remove the six bolts which retain the cam spindle steady plate. Note that each bolt is fitted with a shakeproof washer. Detach the steady plate.
3   The camshafts are interchangeable and as such should be marked before removal, to aid correct replacement. Suitable identification marks may be scribed on each cam wheel face. Withdraw both cams and the Magdyno idler pinion.
4   Lock the crankshaft by placing a close fitting steel bar through the small end eye, so that it rests on two equal size

13.4b Engine timing pinions - general view

wooden blocks placed either side of the connecting rod. Loosen and remove the mainshaft end nut and washer. The engine shaft pinion (half-time pinion) which also incorporates the oil pump drive worm gear, is a tight push fit on the parallel shaft and is secured by a Woodruff key. The pinion should be removed using a small two-legged sprocket puller, the legs of which can be located behind the straight-cut gear teeth. If the sprocket puller is not available, the pinion may be eased off the shaft using suitable levers. If the latter procedure is adopted, special care must be taken not to damage the gear, the casing or the mainshaft. Prise the Woodruff key from position and store it in a safe place to prevent loss.

14 Dismantling the engine: removing the camshaft spindles and tappets

1 Removal of the tappets (cam followers) from the timing chest can take place only after the camshaft spindles have been withdrawn from the crankcase. If the crankcase halves are to be separated in any case, the spindles may be drifted out from inside the casing, after the crankshaft assembly has been removed. Removal of the spindles is facilitated by heating the cases, which can be done more easily after complete dismantling. The right-hand crankcase can then be placed in an oven and heated to a uniform and controlled temperature.

2 If the crankcase halves are not to be separated, the camshaft spindles may be removed by using a simple homemade extractor consisting of a short length of tubing, a large washer and a nut and bolt. The tubing can be placed over the spindle to be removed and the bolt screwed into the specially threaded internal position of the spindle. If the nut is then tightened down on the bolt against the large washer placed against the outer end of tube, the spindle will be pulled from position. The tube must be a minimum of 2 5/8 in long and must have an internal diameter of at least 31/32 in, so that it will fit over the spindle shoulder. The correct thread for the bolt is 5/16 in x 26 TPI, which is a BSC thread.

3 Before removing the spindles, note that on most machines a flat is milled on each spindle shoulder. The flat is positioned upwards so that the shoulder does not interfere with the foot of the cam follower. On some later models a spindle with an unmilled small diameter shoulder is fitted, which eliminates this problem.

4 After removal of the spindles, the tappets can be detached. Loosen and remove the two adjuster nuts from each tappet (where fitted). Unscrew the tappet blocks from the casing top. The tappets are now free.

14.3a Remove tappet adjuster screws

14.4a Cam spindles can be drifted out to allow ...

14.4b ... removal of the cam followers

15.1a Use casing screw to withdraw oil pump spindle dowel

15.1b Pull oil pump spindle from housing

15 Dismantling the engine: removing the oil pump

1   To free the oil pump driveshaft, remove the locating dowel from the recess in the timing chest mating surface. The dowel has an internal thread into which a timing cover screw may be fitted, to aid removal. On some machines, the dowel may be capped by a blanking washer. This will have to be prised from position before the dowel can be withdrawn. Provided that the inlet camshaft spindle has been removed, the oil pump spindle can now be displaced upwards into the timing chest. If the cam spindle is still in place the oil pump will have to be removed to enable withdrawal of the oil pump spindle.

2   Invert the engine so that the oil pump is uppermost and so that the crankcase is resting on the cylinder head retaining long bolts. Note the four bolts which pass into the body of the oil pump. Remove only the two bolts which have washers fitted below their heads. The oil pump can now be lifted from place. The other two bolts should not be removed at this stage, and removed later only if the oil pump requires attention since they hold the oil pump components together.

16 Dismantling the engine: separating the crankcase halves

1   Before separating the crankcase halves, the first test for main bearing wear should be made. Grasp each end of the crankshaft in turn and attempt to move the shafts up and down relative to the crankcase. Any movement in any radial direction indicates that the bearing or bearings on that side require renewal. Even if no play is detected, a visual check of the bearings should be made after separation of the crankcase halves as described in Section 18.

2   Remove the nuts and washers from the crankcase securing bolts and studs. Carefully drift each of the bolts from position, taking care not to damage the threads. Initial separation of the crankcase halves may require the use of a rawhide mallet as jointing compound is invariably used on the mating surfaces. Under no circumstances should screwdrivers or other levers be inserted between the mating faces to aid separation, as damage will almost always result, causing subsequent oil leakage. When the joint has been broken, lift the timing side casing off the flywheel assembly. The flywheel assembly can then be lifted from the drive side casing.

3   When separating the crankcases, it will be found that the roller bearing inner races and cages will remain on the mainshafts

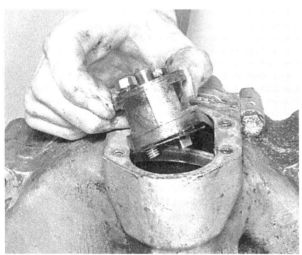

15.2 Oil pump retained by two bolts

16.2a Separate the two casing halves and ...

16.2b ... remove the flywheel assembly

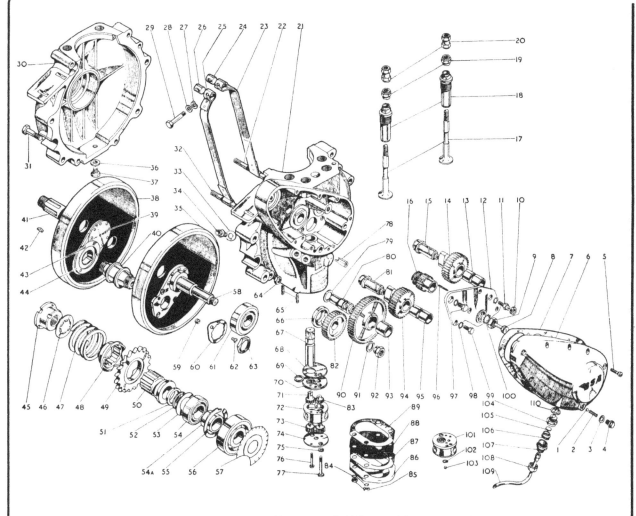

Fig. 1.5. Crankcase assembly - except Gold Star models - component parts

1 Pressure release valve ball	27 Saddle washer	55 Spring ring	82 Magdyno drive pinion
2 Pressure release valve spring	28 Plain washer	56 Drive side roller bearing	83 Feed gear (driven)
3 Pressure release valve fibre washer	29 Bolt	57 Oil thrower	84 Washer - 4 off
4 Pressure release valve screw	30 Drive side crankcase	58 Timing side mainshaft	85 Nut - 4 off
5 Screw - 9 off	31 Bolt - 3 off	59 Grub screw	86 Sump plate
6 Timing cover	32 Stud	60 Lockwasher - 2 off	87 Gasket
7 Timing cover gasket	33 Fibre washer - 2 off	61 Timing side roller bearing	88 Filter gauze
8 Mainshaft oil feed pipe	34 Oil pipe adaptor - 2 off	62 Screw - 2 off	89 Gasket
9 Half time pinion nut	35 Timing side flywheel	63 Crankpin nut - 2 off	90 Magdyno drive pinion
10 Rubber washer - 2 off	36 Fibre washer	64 Nut - 5 off	91 Spring washer
11 Cam pinion spindle bolt - 2 off	37 Drain plug	65 Stud - 4 off	92 Nut
12 Timing gear plate	38 Drive side flywheel	66 Oil seal	93 Bush
13 Exhaust cam pinion bush - 2 off	39 Rivet - 14 off	67 Oil pump drive spindle	94 Inlet cam pinion
14 Exhaust cam pinion	40 Crankpin	68 Gasket	95 Bush - 2 off
15 Exhaust cam pinion spindle	41 Drive side mainshaft	69 Top cover washer	96 Woodruff key
16 Half time pinion	42 Woodruff key	70 Top cover	97 Bolt
17 Tappet - 2 off	43 Rivet - 6 off	71 Feed gear (driven)	98 Washer - 6 off
18 Tappet guide - 2 off	44 Thrust washer - 2 off	72 Oil pump body	99 Bolt - 3 off
19 Tappet locknut - 2 off	45 Cush drive nut	73 Oil pump scavenge gear - 2 off	100 Washer
20 Tappet head - 2 off	46 Cush drive lockwasher	74 Oil pump bottom cover	101 Oil pump
21 Timing side crankcase	47 Cush drive spring	75 Spring washer - 2 off	102 Circlip
22 Stud	48 Cush drive sleeve	76 Bolt - 2 off	103 Ball
23 Magdyno strap (short)	49 Engine sprocket	77 Bolt - 2 off	104 Breather union
24 Tapped plug	50 Cush drive bearing	78 Retaining pin	105 Disc
25 Magdyno strap (long)	51 Cush drive spacer	79 Thrust washer	106 Screw
26 Plain plug	52 Cush drive shim	80 Spindle	107 Breather body
	53 Mainshaft shim	81 Spindle	108 Gland nut
	54 Drive side ball bearing		109 Breather pipe
	54a Mainshaft distance collar		110 Fibre washer

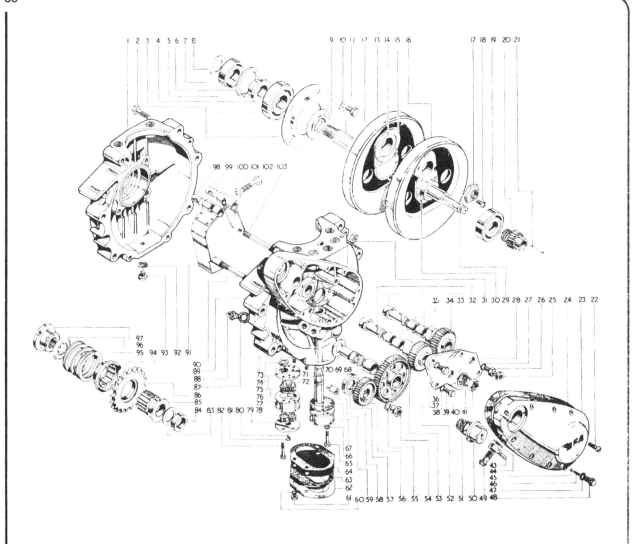

Fig. 1.6. Crankcase assembly - Gold Star models - component parts

1	Drive side crankcase	28	Timing gear plate	52	Inlet cam pinion	77	Oil pump body
2	Bolt - 3 off	29	Mainshaft (timing side)	53	Idler pinion	78	Oil pump scavenge gear - 2 off
3	Bearing retainer	30	Rivet - 14 off	54	Magdyno drive nut	79	Oil pump bottom cover
4	Drive side roller bearing	31	Nut - 5 off	55	Spring washer	80	Oil pump valve ball
5	Spacer	32	Cam pinion spindle (inlet)	56	Magdyno drive nut	81	Circlip
6	Locating ring	33	Cam pinion spindle (exhaust)	57	Oil seal	82	Bolt - 2 off
7	Drive side ball bearing	34	Exhaust cam pinion	58	Retaining pin	83	Cush drive shim
8	Mainshaft shim	35	Cam pinion bush - 4 off	59	Gasket	84	Cush drive distance piece
9	Spacer	36	Bolt - 1 off	60	Washer - 4 off	85	Bearing
10	Lockwasher - 4 off	37	Washer - 3 off	61	Nut - 4 off	86	Engine sprocket
11	Bolt - 4 off	38	Washer - 3 off	62	Sump plate	87	Fibre washer - 2 off
12	Mainshaft (drive side)	39	Bolt - 3 off	63	Gasket - 2 off	88	Oil pipe adaptor - 2 off
13	Flywheel	40	Washer	64	Filter gauze	89	Timing side crankcase
14	Rivet - 6 off	41	Timing pinion nut	65	Bolt - 2 off	90	Stud
15	Thrust washer - 2 off	42	Timing pinion nut	66	Spring washer - 2 off	91	Magdyno strap (front)
16	Flywheel	43	Breather union	67	Oil pump	92	Fibre washer
17	Crankpin nut - 2 off	44	Fibre washer	68	Idler pinion bush	93	Drain plug
18	Screw - 2 off	45	Pressure release valve ball	69	Idler pinion spindle	94	Cush drive sleeve
19	Timing side roller bearing	46	Pressure release valve spring	70	Thrust washer	95	Cush drive spring
20	Half time pinion	47	Pressure release valve fibre washer	71	Oil pump drive spindle	96	Lockwasher
21	Woodruff key			72	Oil pump feed gear (driven)	97	Cush drive nut
22	Screw - 9 off	48	Pressure release valve retaining screw			98	Tapped plug
23	Timing cover			73	Stud - 4 off	99	Plain plug
24	Gasket - 2 off	49	Breather banjo bolt	74	Washer	100	Magdyno strap (back)
25	Rubber washer - 2 off	50	Engine breather valve	75	Oil pump top cover	101	Washer
26	Bolt - 2 off	51	Engine breather valve spring	76	Oil pump feed gear (driven)	102	Bolt
27	Washer					103	Stud

but the journal ball bearing and roller bearing outer races will remain in the cases.

17 Examination and renovation: general

1   Before examining the parts of the dismantled engine unit for wear, it is essential that they should be thoroughly cleaned. Use a petrol/paraffin mix to remove all traces of old oil and sludge that may have accumulated within the engine and a cleansing compound such as 'Gunk' or 'Jizer' for the exterior surfaces of castings. Special care should be taken when using these latter compounds, which require a water wash after they have worked into the film of grease and oil. Water must not be allowed to enter any of the internal oilways or electrical parts such as the dynamo or magneto.

2   Examine the crankcase and rocker box castings for cracks or other signs of damage. If a crack is discovered, it will require specialist repair.

3   Examine carefully each part to determine the extent of wear, if necessary checking with the tolerance figures listed in the Specifications Section of this Chapter.

4   Always use a clean, lint-free rag for cleaning and drying the various components prior to reassembly, otherwise there is risk of small particles obstructing the internal oilways.

18 Main bearing and flywheel assembly: examination and renovation

1   If any up and down movement of the flywheel assembly was experienced when carrying out the test described in Section 16.1, no further investigation of the main bearings is required. They should be renewed as a set, as a matter of course.

2   Check the outer races of the roller bearings for 'shadowing', pitting or flaking of the hardened surface and check the journal ball bearing(s) for roughness during rotation or pitting of the tracks. The bearings should be absolutely clean before inspection takes place.

3   The roller bearing outer races and the ball bearings may be drifted from the cases from the outside, after the crankcase halves have been heated. Heating can best be done in an oven to a temperature of approximately 100°C (212°F). If an oven is not available, immersion in boiling water is a good substitute. On Gold Star models it is necessary to remove the securing plate retained by four bolts, before the drive side bearings can be removed.

4   On all models a spacer is fitted between the two drive side bearings. The drive side ball bearing is retained in position by a large circlip, which must be removed before any attempt to drift out the bearing is made.

5   The roller bearing inner races and cages can be removed along the mainshafts by the judicious use of tyre or other levers placed behind the inner race or oil thrower plate. Gold Star models are not fitted with an oil thrower plate.

6   Play in the main bearings, particularly on the timing side, will almost certainly lead to wear in the bush fitted to the cam spindle support plate. This bush gives support to the outer end of the mainshaft via the boss on the engine shaft pinion.

7   Check the big-end bearing for wear by pulling and pushing the connecting rod in a vertical plane. There should be no play whatsoever if the bearing is to continue in service. The correct side clearance of the connecting rod should be 0.010 - 0.012 in (0.25 - 0.3 mm). Renewal of the big-end bearing requires some experience and the use of various measuring instruments to ensure the requisite accuracy in re-aligning the flywheels. It is therefore recommended that the complete flywheel assembly be entrusted to a competent motorcycle repair specialist for the work to be carried out. The rebuilt assembly must be realigned with a lathe.

8   New big-end bearings are usually available as an assembly, complete with an exchange connecting rod, in which case the complete unit will be ready for refitting. In other cases the

18.2 Check the bearing outer races for shadowing

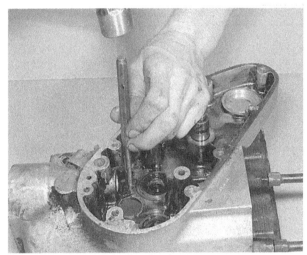
18.3 Drift races out only after heating cases

18.5 Prise bearings from mainshafts

bearing is supplied separately. In the latter instance the outer race of the old bearing will require pressing from the connecting rod and the new outer race pressed in. The new race will probably require honing so that the rollers are a perfect fit.

9   Provided that the roller track of the crankpin and that of the outer race is not scored or pitted it is permissible to take up a small amount of wear by honing the outer race and fitting slightly oversize rollers. This method of bearing reclamation is only recommended as a last resort and should not be applied to a competition machine where the stress is consistently greater than that encountered by a road going machine.

10 When realigning the flywheel assembly the maximum permissible run-out is 0.002 in at the mainshafts.

11 Before refitting the flywheel assembly, check that the locking plates are fitted correctly on the crankpin nuts and that the screws are tight and secured either by caulking or a locking fluid. This also applies to the blanking grub screw in the end of the crankpin.

12 A worn big-end bearing can be recognised by a characteristic knock and accompanying vibration. Do not use the machine in this condition since there is grave risk of a broken connecting rod and accompanying severe engine damage, especially if the rod fractures whilst the engine is highly stressed.

19 Small end bush: examination and renovation

1   The small end bearing of the connecting rod takes the form of a phosphor bronze bush with an oil hole and an internal oil groove. The gudgeon pin should be a good sliding fit in the bush, without evidence of any vertical play. When play develops, the bush must be removed and renewed, but it is first advisable to check whether the wear has not occurred in the gudgeon pin itself.

2   A simple extractor can be made to remove the old bush and at the same time pull the new bush into location. Refer to the accompanying diagram. It must be aligned so that the oil hole registers exactly with the slot in the top of the connecting rod, or the bearing will be starved of oil. When in the correct position, it must be reamed out so that the gudgeon pin is once again a good sliding fit, with a complete absence of vertical play.

20 Timing pinions: examination and renovation

1   It is unlikely that any of the timing pinions will require attention unless the machine has covered an exceptionally high mileage or if the entry of some foreign body has caused accidental damage. Check the pinions for damage or chipped teeth and reject any that are damaged in this respect. Note that each pinion has timing marks in the form of a dot and/or a dash, to facilitate ease of valve timing during engine reassembly.

2   After considerable service the bronze bushes upon which the two cam wheels and the idler pinion run may become worn. The cam wheels are fitted with two bushes each, whereas the idler pinion has only one. The old bushes may be pushed from position using a suitable mandrel, the outside diameter of which is fractionally smaller than that of the bush in question. Use a large vice to apply the necessary pressure. New bushes may be refitted, using the same technique.

3   After refitting the bushes, check that each pinion is an easy sliding fit on its spindle. Tight bushes will require easing with a reamer before final assembly. Do not attempt to fit a pinion with a tight bush in the hope that it will run in. The bronze material used in the bushes is prone to pick-up and seize at the slightest provocation.

4   On final assembly the endfloat between the pinions and support plate should be 0.002 - 0.003 in (0.050 - 0.075 mm).

5   When checking the cam wheels for damage also inspect the cam lobes. The cams and the cam followers in this engine are not prone to excessive wear, due to the relatively large dimensions of the rubbing area and to the generous supply of lubricant.

6   Wear of the cams can usually be detected in the form of scuff marks, break through of the hardened surfaces, or in an extreme case, a ridged and rippled surface. Provided the case hardened surface of the cams has not broken through, it is permissible to remove any slight imperfections by the use of an oilstone. If there is danger of changing the cam profile or reducing the lift, a new cam wheel should be fitted without question.

7   If the cams are worn, the cam followers will show signs of wear too and the two complete assemblies should be renewed at the same time. Otherwise examine the followers for the same forms of imperfection; if there is any doubt about their condition, they should be renewed.

21 Cylinder barrel: examination and renovation

1   Unless the cylinder barrel is new or has recently been rebored, there will be a lip at the top of the bore that denotes the limit of travel of the top piston ring. The depth of this lip will give some indication of the extent to which wear has taken place, even though the amount of wear is not evenly distributed.

2   Remove the rings from the piston, taking great care as they are brittle and very easily broken. Most wear occurs within the top half of the bore, so the piston (without rings) should be inserted in this area and the clearance between the skirt and the cylinder wall measured. The correct clearance for each model is given in the Specifications at the beginning of this Chapter. Clearance slightly more than recommended may be overlooked if in all other respects the bore is in good condition. If the clearance is excessive, a rebore will be required, and a new oversize piston fitted. Similar attention will also be required if the cylinder bore is scored or damaged in any way. A displaced circlip will give rise to a vertical groove, which can be responsible for compression loss if it is not eliminated. Oversize pistons are available in + 0.020 in and + 0.040 in sizes for Gold Star models and + 0.020, + 0.030, + 0.040 and + 0.060 in for all other models.

3   When a rebore is to be carried out, first acquire a piston of the correct oversize so that the cylinder barrel may be bored out to suit the piston exactly. This will ensure that an accurate and long lasting job results.

4   Check that the outside of the cylinder barrel is clean and free from road dirt. Use a wire brush on the cylinder fins if they are obstructed in any way.

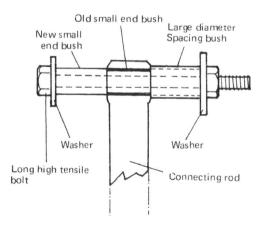

Fig. 1.7. Using drawbolt method to renew small-end bush

5 Never omit the compression plates under the cylinder barrel when the engine is reassembled. Apart from the danger of the valves hitting the piston when they are on full lift, if the cylinder barrel has a lip at the top it will be contacted by the top piston ring, giving rise to a broken ring or even a cracked piston. This advice applies only to Gold Star machines, some of which were fitted with compression plates as standard to modify the compression ratio.

22 Piston and piston rings: examination and renovation

1 Attention to the piston and piston rings can be overlooked if a rebore is necessary, since new replacements will be fitted.
2 If a rebore is not considered necessary, examine the piston carefully. If score marks are evident, or if the skirt is badly discoloured as the result of exhaust gases by-passing the rings, a new piston and rings should be fitted. Check that the correct size is fitted; if the original piston was oversize, the amount will be stamped on the crown.
3 Remove the carbon from the piston crown, using a blunt scraper which will not damage the surface. Clean away all carbon deposits from the valve cutaways and finish off with metal polish so that a clean, shining surface is achieved. Carbon will not adhere so readily to a polished surface.
4 Check that the gudgeon pin bosses are not worn or the circlip grooves damaged. Check that the piston ring grooves are not enlarged. Side float must not exceed the amount laid down in the Specifications.
5 Piston ring wear can be measured by inserting each ring in the cylinder bore from the top and pushing it down with the base of the piston until it is square in the bore about 1½ inches from the top. The figures given in the Specifications are for newly fitted rings. On Gold Star models a small increase may be overlooked. On all other engines a maximum of 0.024 in (0.60 mm) is permissible. In addition to the end gap when fitted, the rings must retain a certain amount of springiness. Measure the end gap of each ring in the free state. If this is less than approximately 3/16 inch the rings should be renewed as a set.
6 When fitting new rings, the end gaps must be checked to ensure that sufficient clearance is available. If the gap is too small, the ring ends may butt together, when the engine is running, causing ring breakage and seizure. If necessary, the gap may be enlarged by carefully placing one end of the ring in the soft jaws of a vice so that the end is firmly held. A very sharp swiss file or small carborundum stone can then be used to remove the required amount of metal. The rings are very brittle and consequently great care must be taken to avoid fracture. The need for a rebore cannot be obviated by fitting oversize rings, even temporarily. This practice should be avoided at all costs. It will lead to reduced compression and a much higher rate of oil consumption.
7 When fitting new piston rings, or if the engine has seen long service, a check should be made to ensure there is no build up of carbon behind the rings, or in the grooves of the piston. Any build up should be removed by gentle scraping.
8 Do not omit to examine the gudgeon pin. Wear is most likely to occur where the pin passes through the small end bush and can be measured with a micrometer. If there is play at the small end bush and the bush itself is in good order, the gudgeon pin should be renewed.

23 Valves, valve springs and valve guides: examination and renovation

1 Before the valves, valve springs and valve guides can be examined, it is necessary to remove the valves from the cylinder head. This is accomplished by means of a valve spring compressor, which will compress each spring sufficiently to

23.1a Two valves springs are fitted to each valve

23.1b Do not omit valve spring seat on reassembly

23.1c Collets must be securely located in collar

23.1d Note valve stem cap fitted to some models

23.9 Valve guide may fracture during removal

permit the split collets to be removed from each valve stem. When the collets have been detached the compressor can be unscrewed until the valve spring tension is released. The valve, valve spring and collar can then be removed, leaving the valve guide in the cylinder head. To avoid confusion during reassembly do not allow the valves and associated components to become mixed as some of the parts are different.

2   After cleaning the valves to remove all traces of carbon and burnt oil, examine the heads for signs of pitting and burning. Examine the valve seats in the cylinder head. The exhaust valves and their seats will probably require the most attention because they are the hotter running. If the pitting is slight, the marks can be removed by grinding the seats and the valve heads together, using fine valve grinding compound.

3   Valve grinding is a simple, if somewhat laborious task, carried out as follows. Smear a trace of fine valve grinding compound (carborundum paste) on the seat face and apply a suction grinding tool to the head of the valve. Oil the stem of the valve and insert it in the guide until it seats in the grinding compound. Using a semi-rotary motion, grind-in the valve head to its seat, using a backward and forward motion. It is advisable to lift the valve occasionally, to distribute the grinding compound more evenly. Repeat this application until an unbroken ring of light grey matt finish is obtained on both valve

and seat. This denotes the grinding operation is now complete. Before passing to the next valve, make sure that all traces of the valve grinding compound have been removed from both the valve and its seat and that none has entered the valve guide. If this precaution is not observed, rapid wear will take place due to the highly abrasive nature of the carborundum base.

4   When deep pits are encountered, it will be necessary to use a valve refacing machine and a valve seat cutter, set to an angle of 45°. Never resort to excessive grinding because this will only pocket the valves in the head and lead to reduced engine efficiency. If there is any doubt about the condition of a valve, fit a new one.

5   Examine the condition of the valve collets and the groove on the valve stem in which they seat. If there is any signs of damage, new parts should be fitted. Check that the valve spring collar is not cracked. If the collets work loose or the collar splits whilst the engine is running, a valve could drop in and cause extensive damage. Check the condition of the valve stem caps. Renew the caps if indentation is extreme (where fitted).

6   Although the above remarks refer specifically to the overhead-valve models, the procedure is fundamentally the same when working on the M20 and M21 side-valve machines. Before the exhaust valve and spring are removed, the exhaust lifter mechanism should be unscrewed from the forward edge of the tappet chest to improve access.

7   Before grinding in the valves is carried out the clearance between each valve stem and its guide should be checked. A good indication of worn valve guides, is the presence of bluish smoke in the exhaust on over-run. This is caused by oil being drawn past the valve stem by the depression in the inlet tract, and subsequently passing into the combustion chamber and burning.

8   Excessive play between a valve and guide will cause high oil consumption and oiling up of the cylinder, due to oil finding its way down the valve stems. Low compression may also be encountered due to tilting of the valve faces at the seat. A new guide may be fitted to an old valve and vice versa, provided that the old component being used is in good condition.

9   The valve guides are a tight drive fit in the cylinder head, (M20 and M21 valve guides are fitted in the cylinder barrel) and may be drifted from position, using a drift of suitable proportions. If at all possible, a double diameter drift should be used. There is no need to heat the cylinder head, although where an aluminium component is fitted, this will facilitate removal. Take care when removing cast-iron valve guides as they are prone to fracturing, thereby making final removal more difficult. On the exhaust valve guide it is worthwhile removing the build up of carbon from the outer diameter before trying to drift the guide from position.

10 When refitting new guides, it should be noted that on all but side-valve engines, the guides are located by a shoulder on the spring side of the cylinder head. The valve guides fitted to side-valve engines do not have a locating shoulder and must be positioned correctly by measurement. Drive the guides into place from the top of the cylinder barrel until the top of the inlet guide is 1 1/8 in (28.57 mm) and the top of the exhaust guide is 15/16 in (23.8 mm) below the cylinder barrel upper mating surface.

11 Whenever new valve guides are fitted, the valve seats must be re-cut with a 45° seat cutter, to ensure that the valve face is perfectly concentric with the seat. The special cutting tool is of the same type used for many car engines and therefore most garages are able to undertake the work.

12 Check the length of the springs against the figures given in the Specifications given at the beginning of the Chapter. If any spring is substantially shorter than the standard length, it must be renewed. In practice, the valve springs will wear at a similar rate, requiring renewal of the springs as a set.

24 Cylinder head: examination and renovation

1   Remove all traces of carbon from the cylinder head and valve

ports, using a soft scraper. Extreme care should be taken to
ensure the combustion chambers and valve seats are not marked
in any way, especially when an alloy head is being cleaned up.
Finish by polishing the combustion chambers with metal polish
so that carbon does not adhere so easily. Never use emery
cloth since the abrasive particles will become embedded in
the metal, especially in the case of alloy components.
2   Check to make sure the valve guides are free from oil or other
foreign matter that may cause the valves to stick.
3   Make sure the cylinder head fins are not clogged with oil or
road dirt, otherwise the engine may overheat. If necessary, use a
wire brush. When the fins and outer surfaces are clean, a coating
of matt cylinder black will help improve the heat radiation of
cast iron cylinder heads.
4   Reassemble the cylinder head by replacing the valves after
oiling their stems. Compress each set of valve springs in turn,
making sure the split collets are located correctly before the
compressor is released. A light tap on the end of each valve
stem, after reassembly, will act as a double check.

24.2a Remove the spindle acorn nut which ...

24.2b ... allows the spindle to be drifted out (not GS models)

24.3a Note sequence of washers on ...

24.3b ... rocker spindle when refitting

24.4 Use new gasket on rocker covers

25 Rockers, rocker spindles and rocker covers: examination and renovation (all ohv models)

1 It is unlikely that excessive wear will occur in the rockers and rocker spindles unless the flow of oil has been restricted or if the machine has covered a very high mileage. A clicking noise from the vicinity of the rocker boxes is the usual symptom of advanced wear. This should not be confused with the somewhat similar noise that results from excessive tappet clearances.
2 If any shake is present and the rocker arm is particularly slack on its spindle, a new rocker arm and/or spindle should be fitted. The spindles are removed by unscrewing the domed nuts from the end of each spindle and tapping them out from the end which is threaded externally, using a brass or copper drift. If the spindles are particularly tight, warming the rocker box casting will aid their removal. On Gold Star and Competition models, the rocker spindles **must** be unscrewed by placing a spanner on the hexagon at the oil union end.
3 Carefully note the sequence of washers and the position of the spindle spring before removing the spindles.
4 Check the cup of each rocker arm where it bears on the push rod and also the valve clearance adjuster on the opposite end (where fitted). If signs of cracking, scuffing or break-through of the case-hardened surface are evident the rocker arm and/or valve clearance adjuster must be renewed. Check also the condition of the thread on the valve clearance adjuster, the thread within the rocker arm, and that of the locknut.
5 Check each pushrod for straightness by rolling on a flat surface. Bent pushrods (usually a sign of over-revving the engine) should be renewed without question since it is not practicable to straighten them effectively.
6 When reassembling the rocker box(es) oil the rocker spindles thoroughly before the rocker arms are replaced and check that the internal oilways are free by pumping oil down the centre of the spindle from the banjo union end, using a pressure oil can.

26 Engine reassembly: general

1 Before reassembly is commenced, the various engine components should be thoroughly clean and placed close to the working area.
2 Make sure all traces of the old gaskets have been removed and the mating surfaces are clean and undamaged. One of the best ways to remove old gasket cement is to apply a rag soaked in methylated spirit. This acts as a solvent and will ensure the cement is removed without resort to scraping, with the consequent risk of damage. More stubborn deposits may be removed using a soft brass wire brush of the type used for cleaning suede shoes, to avoid scratching the surfaces. Because pre-unit motorcycle engines are prone to oil leakage, extra care should be taken when refitting casings. This is particularly so where an engine has been dismantled and reassembled a large number of times, as is consistent with machines of this age. Always use gaskets of the correct material, and where considered necessary, apply a suitable gasket compound to one or both mating surfaces.
3 Gather together all the necessary tools and have available an oil can filled with clean engine oil. Make sure all the new gaskets and oil seals are to hand; nothing is more infuriating than having to stop in the middle of a reassembly sequence because a vital gasket or replacement part has been overlooked.
4 Make sure the reassembly area is clean and that there is adequate working space. Refer to the torque and clearance settings wherever they are given, many of the smaller bolts are easily sheared if they are over-tightened. Always use the correct size spanner and screwdriver, never a wrench or grip as a substitute. If some of the nuts and bolts that have to be replaced were damaged during the dismantling operation, it is advisable to renew them. This will make any subsequent reassembly and dismantling much easier.

27 Engine reassembly: joining the crankcases

1 Before the crankcases are reassembled, check that their mating faces are perfectly clean and free from traces of gasket cement.
2 If the cam followers and spindles were removed or the cylinder head long bolts unscrewed (ohv models only), they should be replaced before proceeding further. Insert the cam followers from inside the timing chest and fit the follower housings after applying a little gasket cement to the threads. Refit the tappet adjuster and locknut (where utilised) to the cam followers. The cam spindles may now be drifted into place. To aid insertion and lessen the risk of damage to the spindle housings, the cases should be heated uniformly to a temperature of approximately 100°C (212°F) in an oven or by means of boiling water. Each spindle should be gently drifted home until the shoulder is hard up against the casing wall. Where the shoulder has a milled flat, the flat must face upwards so that it is parallel with the foot of the cam follower. If this detail is overlooked, the follower will be prevented from moving downwards fully. Each cam spindle is marked for correct location, IN or EX.
3 Screw the cylinder head retaining long bolts into the two crankcase halves. A little gasket compound may be applied to the threads before fitting to prevent possible oil seepage during service.
4 The main bearings may now be fitted into the crankcase and onto the flywheel mainshafts. Proceed by fitting the drive side outer ball bearing and the retaining circlip. Do not omit the large washer, which should be fitted between the bearing and the locating shoulder in the housing. When fitting bearings, the crankcase should be heated as when replacing cam spindles, to allow ease of insertion. Replace the inner main bearing outer race, ensuring that it is hard up against the locating shoulder. On Gold Star models the bearing retaining plate, which is retained by four bolts, should be replaced. Under no circumstances should the tab washers be omitted, the ears of which must be bent up to secure the bolts. A small amount of locking fluid may be applied to the threads of each bolt, as an added precaution against loosening.
5 Replace the main bearing(s) in the timing side crankcase. M20 and M21 models are fitted with a journal ball bearing in addition to the roller bearing. The ball bearing must be inserted first. In all cases the outer race only of the roller bearing should be fitted.
6 Position the oil flinger plate on the drive side mainshaft so that the cut away engages with the crankpin nut. Slide the inner race/roller portion of the drive side inner bearing over the mainshaft and carefully drift it home, using a suitable length of tubing. Fit the timing side bearing in a similar manner.
7 Slide the bearing spacer onto the drive side mainshaft. The flywheel assembly can now be fitted into the drive side crankcase half. Take care when inserting the flywheel assembly that the connecting rod does not foul the crankcase mating face.
8 Apply gasket cement to both faces of the crankcase. The amount used will depend on the condition of the surfaces. Refit the Magdyno (or magneto) retaining straps onto the two studs projecting from the timing side crankcase half. If the straps are of different lengths, the longer one should be placed on the outer stud.
9 Support the drive side crankcase so that the flywheels are uppermost. Position the timing side crankcase half over the main shaft and lower it into position. If necessary, use a rawhide mallet to aid assembly. Fit the crankcase retaining bolts and nuts. Before tightening the nuts, ensure that the upper mating surface of the crankcase halves is absolutely flush. If necessary, rotate the timing side crankcase half until alignment is correct. Before proceeding further, check that the flywheel assembly is free to rotate on the main bearings.
10 It is worthwhile, at this stage, checking that the connecting rod is absolutely central in the crankcase register. Fit the primary drive sprocket components and nut so that the flywheel assembly

27.2 Cam spindles are marked to aid reassembly

27.4a Refit outer bearing and retaining circlip

27.4b Replace outer races only when cases have been heated

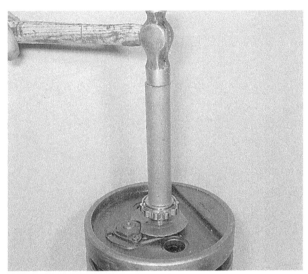

27.4c Drive bearings onto mainshaft

27.7 Do not omit drive side bearing spacer. Before ...

27.8a ... fitting flywheel assembly into drive side casing half ...

is drawn hard up against the drive side inner bearing. The measurement can now be made. If the connecting rod is not central, it will probably be over to the drive side due to wear in the bearing spacer or drive side flywheel cheek. To remedy this problem it will be necessary to separate the crankcases and fit a shim of suitable thickness between the bearing spacer and the inner bearing race.

28 Engine reassembly: replacing the oil pump and crankcase sump plate

1   Invert the completed crankcase unit so that it is resting on the cylinder barrel studs. Insert the oil pump spindle through the sump plate aperture and replace the spindle locating dowel so that it engages with the annular groove in the spindle. If the outer end of the dowel is deeply recessed in the timing chest mating surface, a blanking washer may be fitted to improve location. Under no circumstances should the thickness of the washer be such that the washer stands proud of the mating surface or the dowel will bind against the oil pump spindle when the timing cover is fitted.
2   Apply a small amount of gasket compound to both sides of the oil pump base gasket. Do not use an excessive amount of

compound or it may find its way into the oil passages, causing lubrication failure. Position the oil pump so that the slot in the drive spindle engages with the torque protruding from the oil pump. Fit and tighten evenly the two pump retaining bolts and washers. Rotate the oil pump spindle during tightening, to ensure that it moves freely. If the bolts are overtightened the pump will bind.
3   Place a new gasket on the sump plate studs, followed by the gauze oil trap, the second gasket and the sump plate. Apply gasket compound to both sides of each gasket before fitting. If the four studs are loose, they should be treated with locking fluid before tightening. Fit and tighten the four plate retaining nuts and washers.

29 Engine reassembly: replacing the camshaft and timing pinions

1   Insert the long Woodruff key into the keyway in the timing side mainshaft and slide the half-time pinion onto the shaft so that it engages with the Woodruff key. As the pinion is pushed home the worm gear on the inner end will mesh with the oil pump drive spindle gear. To aid meshing, rock the flywheel assembly and rotate the oil pump spindle by hand to draw the pinion into mesh.

27.8b ... fit Magdyno retaining strap before reassembling cases

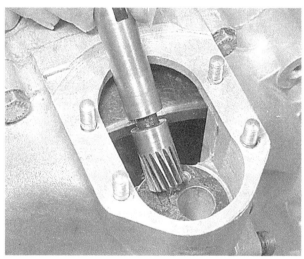

28.1a Insert the oil pump spindle and ...

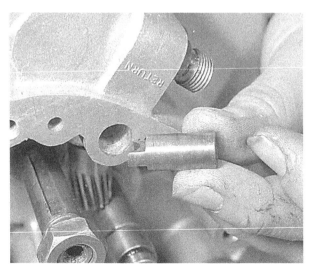

28.1b ... fit the spindle retaining dowel

28.2a Apply compound and fit pump gasket followed by ...

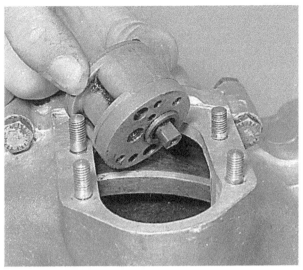

28.2b ... oil pump which is retained by two bolts

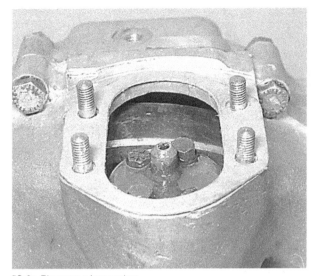

28.3a Fit sump plate gasket ...

28.3b Oil filter gauze screen and ...

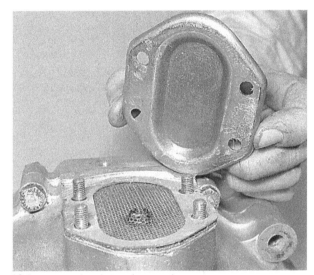

28.3c ... second gasket and sump plate

29.1a Refit Woodruff key into mainshaft and ...

29.1b ... slide half-time pinion into position

2   Place the crankshaft at approximately top dead centre (TDC) so that the two timing marks on the half-time pinion are uppermost. Fit the inlet camshaft so that the dash mark on the cam pinion aligns with the dash mark on the half-time pinion. Fit the exhaust camshaft so that the dot on the cam pinion aligns with that of the half-time pinion, without turning the crankshaft. On all models except some Gold Star machines, the inlet and exhaust camshafts are interchangeable, both being marked with a dot and a dash timing mark on each pinion. When timing remember that the dash mark should only be used for timing the inlet camshaft, and the dot mark only for the exhaust.

3   On most Gold Star models the pinions are similarly marked but the two camshafts are of different types. They should not be interchanged. If there is any doubt as to which camshaft is which, consult the table of camshafts in the Specifications at the beginning of the Chapter. Each cam pinion is marked with the part number. Replace the idler pinion.

4   Replace the cam spindle steady plate and fit and tighten the retaining bolts. Note that the coarse threaded bolts screw into the crankcase bosses. Check that each pinion has the required endfloat of 0.002 - 0.003 in (0.05 - 0.075 mm). Adjustment can be made only by the addition of suitable shims.

5   Lock the crankshaft by placing a close-fitting rod through the small end eye, resting on two wooden blocks placed across the crankcase mouth. Fit and tighten the timing pinion retaining nut.

---

30 Engine reassembly: fitting the piston and cylinder barrel

1   Before fitting the piston, it is advisable to pad out the crankcase mouth with clean rag, in order to prevent a displaced circlip from falling in. Extra, unnecessary dismantling work may be called for if the worst happens and this precaution is not observed.

2   Oil the small end, the piston bosses and the gudgeon pin. Fitting is made easier if the piston is warmed first, especially when a new piston is being fitted after a rebore.

3   As the gudgeon pin is fitted, check that the circlips have engaged with their retaining grooves. A misplaced circlip can work free and cause extensive engine damage, especially if the gudgeon pin is allowed to work out of position and score the cylinder bore.

4   Always use new circlips. It is false economy to use the old components, even if they appear perfect. There is too much to

risk if a circlip breaks or works free whilst the engine is running.

5   The piston should be replaced in the position it was in before removal. A new piston of the flat top type, or one with identical sized valve cutaways may be fitted either way round, unless it is of the split-skirt variety, when incorrect fitment will result in rapid piston and bore wear, and possible engine seizure. Similarly, where the valve cutaways are of differing sizes, the piston must be placed so that the larger cutaway corresponds with the inlet valve, which has a larger head than the exhaust valve.

6   Fit a new cylinder base gasket, and where originally fitted, the compression plate(s). Refitting of compression plates where a new piston is to be used after a rebore is a matter of choice, but where the old piston and bore is to be used, the compression plate(s) must be replaced. If this precaution is not observed, the top piston ring will strike the lip worn previously at the upper limit of top ring travel, and will certainly fracture.

7   Oil the cylinder bore with clean engine oil and arrange the piston rings so that their end gaps are approximately 120° apart on the piston. Rest the piston on two strips of wood placed across the crankcase mouth, then lower the cylinder barrel into position. Before fitting the cylinder barrel on Gold Star models, the short bolts, which pass upwards through the fins and retain the cylinder head, should be refitted. To prevent the bolts falling free, loop a length of inner tube or a large elastic band around the lower portion of the barrel, to secure the bolts in place. The piston rings may be compressed, one at a time, as the lowering of the barrel is accomplished. There is a tapered lead-in at the base of the cylinder bore that will facilitate fitting. When the oil control ring has entered the bore fully, remove the supporting wood strips followed by the packing rag. Push the cylinder barrel downwards until it seats on the crankcase.

8   On side-valve models, fit the five cylinder barrel retaining nuts and washers and tighten them down evenly, in a diagonal sequence.

---

31 Engine reassembly: replacing the Magdyno (or magneto) and retiming the ignition

All except alternator models

1   Timing the ignition can be carried out with ease if it is accomplished whilst the cylinder head has not been refitted to the engine. In this state the position of the piston in the bore may be readily ascertained and set.

29.2 Camwheels must be aligned as shown

29.4a Fit spindle steady plate and feed bolts and ...

29.4b ... fit and tighten steady plate bolts

29.5 Replace the mainshaft nut

30.2 Insert the gudgeon pin and fit the piston

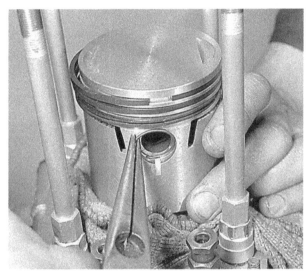
30.3 Ensure that circlips are correctly located

30.7 Slide the cylinder barrel into place

31.3 Fit a new magneto pinion oil seal before ...

31.8 ... timing the ignition and fitting pinion

31.9a Always use a new timing chest gasket

31.9b Lubricate pinions before replacing cover

2   Position the Magdyno on the platform of the crankcase and loosely attach the retaining straps. If shims were fitted between the magneto base and the support, they must be replaced or the mesh between the idler pinion and magneto pinion will be too tight. If a new Magdyno is to be fitted, the mesh between the idler pinion and magneto pinion must be checked by tightening the retaining straps and temporarily refitting the magneto pinion. There should be a small amount of backlash between the teeth. No backlash will cause a mysterious whining noise; too much will cause gear clatter. The adjustment can be made by fitting or removing shims.

3   If a new magneto seal is required, it should be tapped into position after greasing the sealing lip. To fit the Magdyno, push it inwards towards the timing case and tighten the strap bolt.

4   Rotate the magneto armature until the contact breaker points are in the fully open position. Check that the points gap is within the range 0.012 in (0.30 mm) using a feeler gauge. If the points gap is incorrect, adjustment must be made before ignition timing is commenced.

5   Rotate the crankshaft until the piston is at TDC on the compression stroke (with both valves closed). Rotate the engine backwards about half a stroke and then forwards until the piston is the following distance from TDC, still on the compression stroke.

M20 and M21	7/16 inch	BTDC	
B31, B32, B34	7/16 inch	BTDC	
B33 and M33	3/8 inch	BTDC	

Gold Star models

CB32 and DB32	15/32 inch	BTDC	39°
CB34 and DB34			
(except DB34 Touring)	13/32 inch	BTDC	30°
DB34 Touring	1/2 inch	BTDC	41°
DB34 and DB34			
Clubman	15/32 inch	BTDC	39°

6   On all models the ignition timing is given in the advance position ie; when the manual advance-retard cable is in the slack position.

7   TDC on the compression stroke and the correct piston position may be found by supporting a straight-edge across the top of the cylinder barrel from which the measurement can be taken using a finely calibrated ruler or a Vernier gauge. Alternatively, a dial gauge can be mounted on the cylinder barrel. With the exception of side-valve models, it may be necessary to clamp the cylinder barrel down by means of one of the cylinder head long bolts, a nut and a suitable spacer, to take the place of the cylinder head. This will prevent the barrel lifting during the upward movement of the piston.

8   Turn the magneto armature in a clockwise direction (viewed from the contact breaker side) until the points are just on the verge of opening. This can be ascertained by placing a slip of cigarette paper between the points. At the moment the points begin to open they will lessen their grip on the paper. Hold the magneto armature in the correct position and tap the pinion into place so that it engages with the tapered shaft. On machines fitted with a timed breather driven from a peg on the pinion, the pinion must be aligned so that the peg is exactly in line with the timing cover screw hole immediately above the idler pinion (see the accompanying diagram), when the piston is at TDC.

9   Before fitting and tightening the magneto pinion retaining nut, make a check that the timing is accurate. Move the contact breaker cam in the direction of magneto travel to the fully retarded position. Re-insert the cigarette paper and slowly advance the cam. If the timing is correct, the points will begin releasing their grip on the paper at the furthest extent of cam travel.

10   The timing cover can now be replaced. Fit a new gasket and ensure that the oil feed seals are fitted to the inside of the timing cover. On some models the breather valve must be fitted to the magneto pinion before cover replacement.

32 Engine reassembly: refitting the contact breaker assembly and timing the ignition

Alternator models only

1 Position the piston at 7/16 inch (B31) or 3/8 inch (B33) BTDC on the compression stroke, using the method described in the previous Section.

2 Fit the contact breaker drive pinion to the shaft and line up the radial drilling in the pinion boss with that in the shaft. Insert the drive pin and fit the securing clip. Place a new gasket on the contact breaker housing. A few blobs of gasket compound on the gasket will help retain it in position.

3 Check that the low tension terminal on the contact breaker housing is in alignment with the bolt passing through the adjustment bracket and into the aluminium housing. If the terminal does not align, loosen the slotted clamp bolt and turn the complete contact breaker base. Retighten the clamp bolt.

4 Turn the contact breaker cam in a clockwise direction, independently of the driveshaft, until it is in the fully advanced position. Now turn the cam further in the same direction until the points are just on the verge of opening. Release the hold on the cam. Ensuring that the relative positions between the drive pinion and the housing do not alter, insert the complete unit into position so that the drive pinion passes into mesh with the idler pinion. The bracket bolt and low tension terminal must be vertical whilst this is carried out. If the drive pinion does not mesh with the idler pinion, turn it the fraction of a tooth necessary to ensure meshing. Temporarily refit the three aluminium housing retaining screws and nuts and check the timing. Slight inaccuracy may be rectified by loosening the bracket bolt and moving the bracket within the limits of the elongated hole through which the bolt passes.

33 Engine reassembly: replacing the cylinder head

Side-valve models

1 Grease both sides of the cylinder head gasket and position it on the cylinder barrel. A new gasket should be used unless the old one is in perfect condition.

2 Place the head in position and fit the ten retaining bolts and the plain washers. The cylinder head steady bracket must be fitted at some stage but as it will obstruct refitting the engine, it should not be replaced at this juncture.

3 Tighten the bolts down evenly, in the sequence shown, to prevent distortion. Fit the spark plug to prevent the ingress of foreign matter.

4 The valve clearances should be adjusted as described in Section 34 of this Chapter.

B31, B33 and M33 models

1 Before refitting of the cylinder head can be carried out the valves, spring assemblies and rocker arms should be replaced and the rocker covers fitted. In addition, the exhaust lift mechanism must be replaced correctly in relation to the exhaust rocker arm when that cover is fitted (see diagram). When fitting the acorn nuts to the rocker spindles use new copper washers, unless the old components are perfect. Old washers must be annealed before reuse by heating them to a cherry red colour and plunging them into cold water to quench.

2 Insert a new rubber seal into the cylinder head and insert the pushrod tube and gland nut. Do not tighten the nut at this stage. Grease the cylinder head mating surface to aid sealing as no cylinder head gasket is fitted. Place a new pushrod tube base gasket over the studs, either side of the tappets. The use of gasket compound is recommended, the amount depending on the condition of the mating surfaces.

3 Partially raise the cylinder head retaining long bolts and apply graphite grease to the threads. Pay particular attention to the bolt adjacent to the exhaust port. The application of grease will aid future removal.

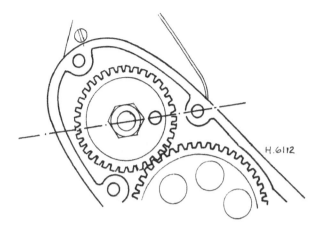

Fig. 1.8. Breather drive peg must align with screw hole when piston is at Top Dead Centre

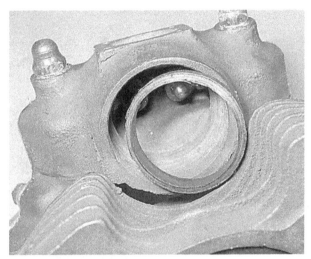

33.2a Fit a new seal to the pushrod tunnel register

33.2b Fit pushrod tunnel and locate pushrods with rockers

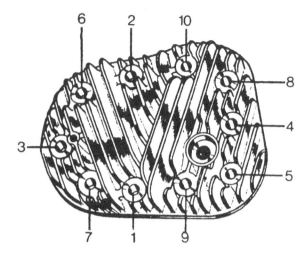

Fig. 1.9. Order of tightening cylinder head bolts
(side valve models)

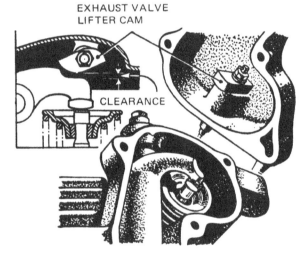

Fig. 1.10. Exhaust valve lifter adjustment

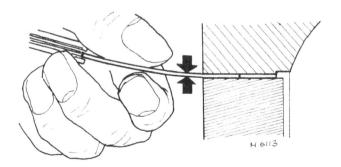

Fig. 1.11. Fitting cylinder head gasket - Gold Star models

4 Insert the pushrods into the pushrod tube and lower the cylinder head down into position on the cylinder barrel. It will be necessary to support the head in a raised position whilst the pushrod ends are engaged with the rocker arm ball ends and the tappets.

5 Fit and tighten the two acorn nuts which retain the base of the pushrod tube, and then tighten the four cylinder head long bolts evenly, in a diagonal sequence. The pushrod tube gland nut can now be tightened sufficiently to compress the rubber ring and effect a gas and oil tight seal.

6 Before replacing the upper and lower inspection plates, the tappet clearances must be adjusted as described in the next Section.

Gold Star and Competition models

7 The cylinder head joint is effected by the rim of the cylinder sleeve mating with the cylinder head. In addition, a cylinder head gasket is fitted to provide a seal at the integral pushrod tunnel and cylinder head. To ensure that the cylinder head seats on the cylinder sleeve and at the same time the correct 'nip' is given to the head gasket to seal the pushrod tunnel, the gasket is laminated in aluminium foil. This allows successive lamination to be peeled from the gasket and so the correct thickness can be arrived at.

8 To ascertain what thickness of gasket must be used, position the cylinder head on the cylinder barrel and tighten the retaining bolts down lightly. Measure the gap between the cylinder head mating surface and that of the cylinder barrel (see diagram) using a feeler gauge. Measure the cylinder head gasket thickness and if necessary peel off a number of laminations until the gasket is 0.001 - 0.002 inch (0.025 - 0.05 mm) thicker than the gap. The laminations are 0.005 inch (0.013 mm) thick.

9 Before the cylinder head is finally fitted to the engine, the rocker box complete with rockers and spindles must be replaced. Although it is physically possible to fit the rocker box last, this should not be done as an uneven strain will be placed on the retaining studs and warpage of the rocker box may occur.

10 Insert the pushrods into the integral tunnel in the cylinder barrel so that their lower ends locate in the recesses in the tappets. Place the cylinder head in position and engage the ball ends of the pushrods with the rocker arms. Carefully tighten the cylinder head bolts evenly, in a diagonal sequence. Fit and tighten the two acorn nuts which retain the cylinder barrel flange on the right-hand side of the engine. As with removal, the two special spanners must be used on the inaccessible bolts.

33.4 Place the complete assembly in position.

34 Engine reassembly: setting the valve clearances

1 Because of the quietening ramps fitted to the cams, the cam followers are only at their lowest point during a relatively small time during the rotation of each cam. Due to this, a special procedure must be adopted when checking and resetting the tappet clearances.

2 The sequence is identical for all models. Turn the engine forwards until the inlet valve has opened and has just closed. At this point the pushrod should be just free to rotate. The engine is now in the correct position for setting the exhaust valve clearance. Turn the engine forwards again very slowly until the exhaust valve clearance has been taken up but before the valve actually begins to lift. The inlet valve may now be set.

Side-valve, B31, B33 and M33 models and Competition models

3 Adjustment of the valve clearance is effected by slackening the locknut below the tappet head and screwing the tappet head upwards or downwards on the threaded tappet. Take the measurement using a feeler gauge inserted between the tappet and pushrod, or in the case of side-valve models, between the tappet and valve stem. After retightening the locknut, check the clearance again.

Tappet clearance	Inlet	Exhaust
M20 and M21 models	0.010 in (0.25 mm)	0.012 in (0.30 mm)
B31, B33 and M33 models	0.003 in (0.08 mm)	0.003 in (0.08 mm)
B32 and B34 Competition models	0.003 in (0.08 mm)	0.003 in (0.08 mm)

Gold Star models

4 Adjustment of the valve clearances is effected by means of eccentric rocker spindles which may be rotated after slackening the acorn nuts, by placing a spanner on the hexagon portion of the spindle at the oil feed end. It should be noted that as the eccentric spindles are turned, the complete rocker assemblies are moved near to or further away from the valves. If adjustment is made when the rocker is furthest away from the valve it operates, the rocker arm pad will not be in correct relationship with the valve stem. To ensure correct positioning of the rockers, rotate the exhaust spindle as far as it will go, without opening the valve, in an anti-clockwise direction. Now move the spindle

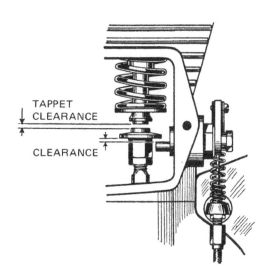

Fig. 1.12. Tappets and exhaust valve lifter (M21)

33.6a Fit the upper pushrod cover followed by ...

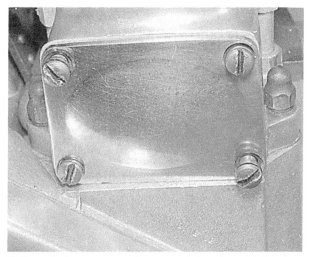

33.6b ... the lower cover, using new gaskets only ...

34.3 ... after setting the valve clearances

35.2 Lift engine into approximately the correct position

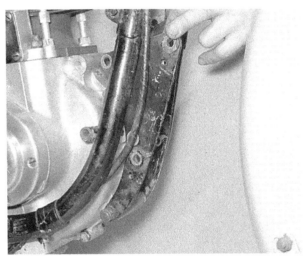

35.3a Fit lower stud first to front engine plates

35.3b Fit all mounting bolts before tightening nuts

back just sufficiently to give the correct clearance. Adjustment of the inlet rocker is similar, except that the spindle must be rotated as far as possible in the clockwise direction. Lock the spindles by means of the acorn nuts and then recheck the clearances.

Valve clearances	Inlet	Exhaust
All models except		
Touring CB and DB Gold	0.006 in	0.006 in
Stars	(0.15 mm)	(0.15 mm)
Touring CB and DB Gold	0.008 in	0.010 in
Stars only	(0.20 mm)	(0.25 mm)

5   Replace the valve gear inspection covers, using new gaskets, and where necessary, a small amount of gasket compound. On side-valve machines, refit the exhaust valve lifter mechanism, ensuring that the lifting peg is positioned below the tappet head.

35 Replacing the engine unit in the frame

1   As when removing the engine, at least two people should be employed in lifting the completed unit into position.
2   Lift the engine into the frame from whichever side is convenient, and allow it to rest in approximately the correct position. Place the two front engine plates either side of the crankcase and insert the mounting bolts. It may be necessary to lift the engine a substantial amount to refit the lowermost mounting stud. Replace the remainder of the studs and fit all the nuts and washers. Do not tighten any nuts until all the studs have been inserted.
3   It should be noted that two long spacers are fitted on each front frame to bracket bolt, and also that two short spacers are fitted on the lower bolt between the frame tubes and brackets. Where the front engine plate cover is retained by the engine mounting bolts, the cover must be replaced as the bolts are fitted. On other models the cover is retained by six small screws and may be replaced after the engine has been fitted.
4   Replace the cylinder head steady bracket, ensuring that the steady plate is not strained into place during tightening down.

36 Engine reassembly: replacing the primary chaincase inner, the primary drive and clutch

1   Apply gasket compound to the steel plate (where utilised) which forms the joint between the rear of the primary drive case inner and the crankcase. Where a Hallite gasket only is used, gasket compound may be used if necessary. Place the steel plate or the gasket in position on the crankcase boss so that the holes align with the threaded holes in the boss.
2   If the gearbox sprocket has been removed, it must be refitted at this stage. Oil the sprocket boss before placing the sprocket onto the splined shaft and pushing it home through the oil seal in the gearbox. The sprocket is retained by a slotted nut and tab washer. The nut should be tightened using a 'C' spanner. If a suitable tool is not available, a brass drift may be employed if care is taken. The sprocket is best prevented from rotation by refitting the final drive chain and applying the rear brake. Do not forget to bend the tab washer up into one of the nut slots in order to secure the nut.
3   Fit the primary chaincase inner and loosely insert the three front retaining bolts and washers. On aluminium alloy chaincases, the rear of the case is retained by an additional bolt passing through a lug welded to the frame. This bolt must be refitted before tightening the front screws. The front screws must be wire locked together after tightening, to positively secure them. Failing this, apply a locking fluid to the threads before fitting.
4   On alternator models commence clutch reassembly by assembling the clutch splined sleeve and outer drum on the workbench and inserting the bearing rollers. Lubricate the rollers

36.3a Apply gasket compound to the sealing plate and ...

36.3b ... refit the primary drive case inner

36.4a Replace the Woodruff key in the gearbox mainshaft

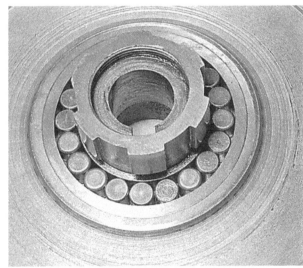

36.4b Assemble the clutch roller bearing on the bench and ...

36.4c ... replace the clutch centre boss before ...

36.4d ... replacing the completed assembly on the shaft

and fit the clutch centre to retain them during further assembly.
Replace the Woodruff key into the recess in the tapered clutch
shaft and position the clutch sub-assembly so that the keyway
in the splined sleeve aligns with the key. Push the assembly
fully home.

5   On all other models replace the Woodruff key and the
splined sleeve, after fitting the oil thrower scroll, and where
utilised, the felt seal and backing plate. Replace the clutch
backing plain plate followed by the bearing inner race, and
the two separate ball cages. The cages must be positioned so that
the flat faces are away from each other. Lubricate the bearings
before fitting the outer drum and clutch centre.

6   Final assembly of the clutch is fundamentally similar for all
models. Fit the backing washer to the clutch shaft (alternator
models only) and replace the tab washer and clutch centre nut.
Tighten the nut fully, preventing the clutch from rotating by
engaging top gear and applying the rear brake. Bend the tab
washer up to lock the nut.

7   Replace the clutch plates one at a time, commencing with a
plain plate and then fitting the friction and plain plates
alternately. Some clutches are fitted with plain plates,
incorporating oil flinger slots. This type of plate should be

replaced so that the slots face in the opposite direction to that
of normal rotation. Before replacing the pressure plate, grease
and insert the operating pushrod into the hollow clutch shaft.
On alternator models a single steel ball is fitted into a recess in
the pressure plate. The ball can be held in position with a dab
of grease during assembly.

8   Place the clutch spring cups and springs in the pressure plate
and start the adjuster nuts. On six spring clutches, nuts and
locknuts are utilised to adjust the springs. On four spring clutches,
special slotted nuts are used, which may be tightened with a
special key. Where two nuts are used, clutch tension is correct
when approximately one thread is visible projecting from the
locknut. If the springs are slightly worn a little more tension
may be required. Adjustment of the four spring clutch is similar
except that the slotted nuts must be screwed inwards until the
heads are flush with the stud ends. Again more tension may be
required if worn springs are being refitted.

9   Check that the plates are lifting squarely by operating the
clutch and depressing the kickstart lever a number of times. If
the pressure plate is out of true, tighten one or two nuts, as
necessary.

10 Replace the shock absorber assembly, including the engine

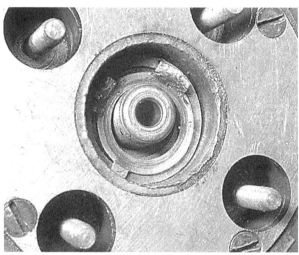

36.6a Replace the tab washer and ...

36.6b ... fit and tighten the centre nut

36.7a Grease and fit the clutch pushrod

36.7b Replace the clutch plates one at a time

sprocket, onto the crankshaft. On alternator models the shock absorber is dispensed with, but the sprocket and backing plate must be fitted. Fit the primary drive chain and connect the ends by means of the spring link, (closed end of link facing direction of travel). A depression in the rear of the chain case allows the spring link to be fitted with ease.

11 The shock absorber sleeve nut can now be tightened fully, using a 'C' spanner or a hammer and punch. It is important that the nut is not allowed to loosen during service. This is particularly important on Gold Star models where the flywheel assembly may disintegrate due to a loose nut. The nut should therefore be tightened very firmly. Bend up the tab washer so that it locates with the recess in the sleeve nut.

12 Turn the engine over until the primary drive chain is in the tightest position and adjust the chain so that there is approximately ½ inch up and down movement in the middle of the upper chain run. Adjustment is effected by slackening the upper and lower gearbox pivot bolts and then turning the drawbolt attached to the right-hand gearbox plate. After adjustment, tighten the locknut on the drawbolt and also the gearbox pivot bolts.

13 On alternator models the alternator should now be replaced, as described in the next Section.

14 Fit a new gasket to the primary chaincase and replace the primary chaincase cover. Gasket compound can be used on both sides of the gasket to ensure an oil tight joint. Models with an aluminium alloy chaincase incorporate the oil level and drain screws in two of the lower chaincase screws, the heads of which were originally painted red for easy identification. The oil level screw is the second screw from the front in the bottom run and the drain screw the fourth screw from the front. Both screws should be fitted with fibre washers, to prevent leakage. The oil level plug on the pressed steel type of chaincase is fitted just below the clutch housing bulge.

15 Refill the chaincase with clean engine oil to the correct level, with the level plug or screw removed. Replace the screw or plug and fit the filler cap. The usual fibre washer is often replaced by a rubber 'O' ring of a suitable size.

37 Engine reassembly: replacing the alternator

1 The alternator may be replaced after refitting the clutch and primary drive as described in the previous Section, paragraphs 1 to 12.

2 Place the Woodruff key in the recess in the crankshaft and slide the alternator rotor into place so that the keyway engages correctly. The rotor must be fitted with the Lucas motif facing outwards. Replace the centre nut and tab washer and then tighten the nut fully. Bend the tab washer up to secure the nut.

3 Place one waisted spacer on each of the three alternator stator retaining studs. Replace the wiring lead grommet into the hole in the centre of the chaincase inner. Position the alternator stator with the wiring leads nearest the inner case, and feed the lead ends through the grommet. Carefully slide the stator over the rotor and onto the retaining studs, whilst simultaneously drawing the leads into position from the rear of the case. Fit and tighten the three nuts and spring washers.

4 The rotor must always be replaced with the leads from the coils nearest the inner case and approximately in the 12 o'clock position. Check that the leads do not foul the primary chain.

38 Engine reassembly: completion

1 Now that the engine unit has been assembled and refitted, the ancillary components can be replaced.

2 Reconnect the two leads to the dynamo, referring to the electrical Chapter if there is any doubt as to their correct positions. On alternator models the leads must also be reconnected and the same advice applies. Additionally, on alternator models the low tension lead to the contact breaker

36.7c Ensure that the steel ball is not omitted before ...

36.8 ... fitting the clutch pressure plate and springs

36.10 Ensure that shock absorber nut is TIGHT

housing must be reconnected. Replace the suppressor cap on the spark plug.

3   Reconnect the oil feed and return pipes to the threaded unions in the crankcase. The return pipe must be fitted to the outer of the two unions. Do not confuse the two pipes or lubrication failure will result. Place a new fibre or copper washer on the oil feed end of each rocker spindle (ohv models only) and fit the rocker feed pipes. Do not overtighten the banjo union bolts as they are prone to distortion and shearing. Reconnect the timing chest breather pipe.

4   Reconnect the clutch cable with the exposed operating arm on the gearbox, after passing the cable across through the frame and through the adjuster lug. Adjust the cable so there is a small amount of free play at the handlebar lever before clutch disengagement commences. Tighten the adjuster locknut.

5   Fit a new paper gasket onto the carburettor mounting studs and replace the Tufnol heat sink gasket. On all models but those with a carburettor flange 'O' ring, a second paper gasket must be fitted over the Tufnol heat sink. Position the carburettor on the studs and fit and tighten the retaining nuts. Do not over-tighten as the flange will distort causing leakage at the joint and in extreme cases distortion of the carburettor body causing throttle slide seizure.

When replacing the carburettor top complete with air and throttle slide assembly, make sure that the needle engages with the needle jet or there is risk of bending the needle. It will also be necessary for the air slide to engage with both the slot in the top of the throttle slide and the cutaway in the jet block, before the throttle slide will drop into position.

Make sure that the throttle slide returns when the twist grip is operated and seats correctly when the throttle is closed. A sticking slide can often be attributed to an accumulation of sludge on the sides of the carburettor mixing chamber, which should be cleaned off with a fine abrasive such as metal polish. Lightly oil the throttle slide and check that no abrasive remains within the mixing chamber.

If the throttle valve does not close completely, the engine will tend to race, a mysterious trouble that can usually be traced to a misplaced throttle cable nipple within the throttle slide. If the nipple does not rest within its cutaway, it will prevent the slide from seating correctly. A small split pin is normally used to keep the nipple in position; this pin is often omitted, particularly if the slide has been changed.

Check that the top of the carburettor is tight and is not cross-threaded. If the top works loose, the throttle will jam open, with disasterous results.

On some Competition carburettors, the choke plunger assembly is remote from the main throttle valve assembly. The plunger should be fitted separately. When replacing the remote float chamber on some Competition carburettors, the chamber height must be set accurately to ensure perfect carburettor performance. Refer to Chapter 4, Section 6.9 for the correct procedure.

6   Reconnect the control cable with the exhaust valve lifter lever. Do not omit the push-off spring. Place the engine on the compression stroke so that the exhaust valve is fully closed and adjust the cable so that there is no contact between the lifting cam and rocker arm (tappet head SV models) when the operating arm is in the resting position. There must be a certain amount of play to prevent contact during normal running.

7   Refit the exhaust system as a unit or by replacing the exhaust pipe followed by the silencer.

8   Place the fuel tank in position and fit and tighten the retaining bolt(s). Do not omit the bridge strap which supports the front of the tank (where fitted). Without the strap the tank will flex, which may lead to leakage. Reconnect the fuel feed pipe(s).

9   Replace the left-hand footrest and the brake pedal. Check the brake action and that the stop lamp switch is operated at the correct time.

10 Check that all the oil unions are tight and that the engine and oil tank drain plugs are tight. Refill the tank with the correct quantity of engine oil. Refer to the Specifications.

11 Some machines are fitted with an air cleaner, which is either connected to the carburettor air intake by a small hose or screwed direct to the intake. In either case the air cleaner should be reconnected or carburation will be affected. On no account run without the air cleaner attached. Re-jetting is necessary under these circumstances because the size of the main jet is reduced when an air cleaner is fitted, to compensate for the more rich mixture that results. On some machines the air filter box is bolted to the battery retainer forward strap.

12 Reconnect the battery and secure the unit with the strap. All machines have a positive (+) earth electrical system.

39 Starting and running the rebuilt engine

1   When the engine starts, remove the oil tank filler cap and check that oil is returning. There may be a time lag before oil commences to emerge from the return pipe because pressure has to build up in the rebuilt engine before circulation is complete, but do not permit the engine to run at low speed for more than a couple of minutes before stopping it and checking the system. To check whether the oil pump is working, slacken the pressure release valve and see whether oil issues from the threads when the engine is re-started.

The exhaust will smoke excessively during the initial start, due to the presence of oil used throughout the reassembly process. This should gradually disperse as the engine settles down.

The return to the oil tank will eventually contain air bubbles because the scavenge pump will have cleared the excess oil content of the crankcase. The scavenge pump has a greater capacity than the feed pump, hence the presence of air when there is little oil to pick up.

Check the engine for leakages at gaskets and pipe unions etc. It is unlikely any will be evident if the engine has been re-assembled correctly, with new gaskets and clean jointing faces. Before taking the machine on the road check that both brakes work effectively and that all controls operate freely.

If the engine has been rebored, or if a number of new parts have been fitted, a certain amount of running-in will be required. Particular care should be taken during the first 100 miles or so, when the engine is most likely to tighten up, if it is overstressed. Commence by making maximum use of the gearbox, so that only a light loading is applied to the engine. Speeds can be worked up gradually until full performance is obtained with increasing mileage.

Do not tamper with the silencer or fit another design unless it is designed specifically for a BSA single. A noisier exhaust does not necessarily mean improved performance; in a great many instances unwarranted modifications or the fitting of an unsuitable design of silencer will have an adverse effect on both performances and petrol consumption.

40 Fault diagnosis: engine

Symptom	Cause	Remedy
Engine will not start	Defective spark plug	Remove plug and lay on cylinder head. Check whether spark occurs when engine is kicked over.

	Dirty or closed contact breaker points	Check condition of points and whether gap is correct.
	Ignition coil defective (coil ignition models only)	Remove HT lead from plug and jam between cylinder head fins, take off C/B cover, switch on ignition and using an insulated screwdriver flick the C/B points open - a fat consistent spark should result.
	Fuel starvation	Check fuel supply. Check to see that the fuel tap is turned on. Check the fuel lines for obstruction.
	Fuel flooding	Remove and dry spark plug.
	Low compression	If the engine can be turned over on the kickstart with less than normal effort, perform a compression test and determine the cause of low compression. (To check compression buy or borrow a test gauge of the type that is screwed or held in the spark plug hole while the engine is kicked over).
Engine runs unevenly	Ignition system fault	Check system as though engine will not start.
	Blowing cylinder head gasket, or bad sealing at joint	Leak should be evident from oil leakage where gas escapes.
	Incorrect ignition timing	Check timing and reset if necessary.
	Incorrect fuel mixture	Adjust carburettor. Remove and clean carburettor jets. Check float level. Check for intake air leaks. Make sure that the carburettor mounting bolts are tight.
Lack of power	Incorrect ignition timing	See 'Ignition timing' in Chapter 1.
	Fault in fuel system	Check system and filler cap vent.
	Blowing head gasket	See above.
High oil consumption	Cylinder barrel in need of rebore and o/s piston	Fit new rings and piston after rebore.
	Oil leaks or air leaks from damaged gaskets or oil seals	Trace source of leak and replace damaged gaskets or seals.
	Oil not returning to tank	Remove oil tank filler cap and check whether oil is flowing from the return pipe whilst the engine is running. If trouble persists check oil pump.
Excessive mechanical noise	Worn cylinder barrel (piston slap)	Rebore and fit o/s piston.
	Worn small end bearing (rattle)	Renew bearing and gudgeon pin.
	Worn big-end bearing (knock)	Fit new big-end bearing.
	Worn main bearings (rumble)	Fit new journal bearings.
Engine overheats and fades	Pre-ignition and/or weak mixture	Check carburettor settings. Check also whether plug grade correct.
	Lubrication failure	Check operation of pump as above, also check for possible oilway/oil pipe blockage.

Chapter 2 Gearbox

Contents

General description 1
Dismantling the gearbox: removing the inner and outer
covers (B series only) 2
Dismantling the gearbox; removing the gear clusters and
selector arms (B series only) 3
Dismantling the gearbox: removing the inner and outer
covers (M series only) 4
Dismantling the gearbox: removing the gear clusters and
selector arms (M series only) 5
Examination and renovation: general 6
Examination and renovation: gear pinions, selector arms
and bearings 7
Examination and renovation: kickstart quadrant, ratchet and
return spring 8
Examination and renovation: gearchange mechanism and
return spring (B series only) 9
Examination and renovation: gearchange mechanism

(M series only) 10
Reassembling the gearbox: replacing the camplate and gear
clusters (B series only) 11
Reassembling the gearbox: refitting the inner cover and
kickstart ratchet (B series only) 12
Reassembling the gearbox: refitting the outer cover
(B series only) 13
Reassembling the gearbox: replacing the gear clusters
and inner cover (M series only) 14
Reassembling the gearbox: replacing the gear change mechanism
and outer cover (M series only) 15
Fitting a reverse camplate (Gold Star models only) 16
Changing sprocket sizes 17
Primary chain: lubrication and adjustment 18
Speedometer drive 19
Fault diagnosis 20

Specifications

BSA model									M20	M21	M33
Sprocket sizes											
Engine (solo)	...	...	...	...	...	...	...	...	18	20	20
(sidecar)	...	...	...	...	...	...	...	...	16	16	17
Clutch	...	...	...	...	...	...	...	...	43	43	43
Gearbox sprocket	...	...	...	...	...	...	...	19	19	19
Rear wheel	...	...	...	...	...	...	...	...	42	42	42
Gear ratios (solo)											
Top gear	...	...	...	...	...	...	...	...	5.28 : 1	4.75 : 1	4.75 : 1
3rd gear	...	...	...	...	...	...	...	...	6.95 : 1	6.25 : 1	6.25 : 1
2nd gear	...	...	...	...	...	...	...	...	10.87 : 1	9.77 : 1	9.77 : 1
1st gear	...	...	...	...	...	...	...	...	15.76 : 1	14.15 : 1	14.15 : 1
Gear ratios (sidecar)											
Top gear	...	...	...	...	...	...	...	...	5.94 : 1	5.94 : 1	5.59 : 1
3rd gear	...	...	...	...	...	...	...	...	7.82 : 1	7.82 : 1	7.37 : 1
2nd gear	...	...	...	...	...	...	...	...	12.2 : 1	12.2 : 1	11.5 : 1
1st gear	...	...	...	...	...	...	...	...	17.7 : 1	17.7 : 1	16.75 : 1
Primary chain											
Size	...	...	...	...	...	...	...	...	½ in x 0.305 in	½ in x 0.305 in	½ in x 0.305 in
No. of pitches (solo)	...	...	...	...	...	...	...	69	70	70
(sidecar)	...	...	...	...	...	...	...	...	68	68	69

BSA model									B31	B33
Sprocket sizes										
Engine	...	...	...	...	...	...	...	...	17	19
Clutch	...	...	...	...	...	...	...	...	43	43
Gearbox	...	...	...	...	...	...	...	...	19	19
Rear wheel	...	...	...	...	...	...	...	...	42	42

Gear ratios

Top gear	5.6 : 1	5.0 : 1
3rd gear	6.77 : 1	6.05 : 1
2nd gear	9.86 : 1	8.77 : 1
1st gear	14.42 : 1	12.90 : 1

Primary chain

Size	½ in x 0.305 in	½ in x 0.305 in
No. of pitches	69	70

BSA model	B32	B34

Sprocket sizes

Engine	16	20
Clutch	43	43
Gearbox	16	16
Rear wheel	42	42

Gear ratios

Top gear	7.06 : 1	5.64 : 1
3rd gear	10.22 : 1	8.17 : 1
2nd gear	16.42 : 1	13.14 : 1
1st gear	22.27 : 1	17.82 : 1

Primary chain

Size	½ in x 0.305 in	½ in x 0.305 in
No. of pitches	68	70

Sprocket sizes

B32 Gold Star

	Engine	Gearbox	Clutch	Rear wheel
Touring	17	19	43	42
Scrambles	16	16	43	46
Racing	18	19	43	42
Clubman's	18	19	43	42

B34 Gold Star

	Engine	Gearbox	Clutch	Rear wheel
Touring	19	19	43	42
Scrambles	17	16	43	46
Racing	21	19	43	42
Clubman's	21	19	43	42

Internal gear ratios

	Touring	Scrambles	Racing	Clubman's
Top gear	1.0	1.0	1.0	1.0
3rd gear	1.210	1.325	1.099	1.099
2nd gear	1.758	1.754	1.326	1.326
1st gear	2.580	2.343	1.929	*1.929

*1956 models had 1.754, 1st gear ratio

**Final drive ratios

B32 Gold Star

Top gear	5.6 : 1	7.73 : 1	5.28 : 1	5.28 : 1
3rd gear	6.77 : 1	10.24 : 1	5.8 : 1	5.8 : 1
2nd gear	9.86 : 1	13.56 : 1	7.0 : 1	7.0 : 1
1st gear	14.42 : 1	18.11 : 1	10.15 : 1	10.15: 1

B34 Gold Star

Top gear	5.0 : 1	7.26 : 1	4.52 : 1	4.52 : 1
3rd gear	6.05 : 1	9.62 : 1	4.96 : 1	4.96 : 1
2nd gear	8.79 : 1	12.73 : 1	5.99 : 1	5.99 : 1
1st gear	12.90 : 1	17.0 : 1	8.71 : 1	8.71 : 1

**The Final drive ratios given are the usual standard ratios of machines when despatched from the factory. The following table gives the top gear overall ratio when using permutations of available engine and gearbox sprockets with a standard 43 tooth clutch sprocket.

42 tooth rear wheel sprocket

	Gearbox sprocket	
Engine sprocket	16	19
16	7.05 : 1	5.93 : 1
17	6.63 : 1	5.6 : 1
18	6.26 : 1	5.28 : 1
19	5.93 : 1	5.0 : 1
20	5.64 : 1	4.75 : 1
21	5.38 : 1	4.52 : 1
22	5.13 : 1	4.32 : 1

46 tooth rear wheel sprocket

Engine sprocket												
16 ...	...	...	...	...	...	...	...	...	...	7.73 : 1		6.50 : 1
17 ...	...	...	...	...	...	...	...	...	...	7.26 : 1		6.12 : 1
18 ...	...	...	...	...	...	...	...	...	...	6.85 : 1		5.78 : 1
19 ...	...	...	...	...	...	...	...	...	...	6.50 : 1		5.48 : 1
20 ...	...	...	...	...	...	...	...	...	...	6.18 : 1		5.20 : 1
21 ...	...	...	...	...	...	...	...	...	...	5.89 : 1		4.96 : 1
22 ...	...	...	...	...	...	...	...	...	...	5.6 : 1		4.73 : 1

Primary chain
B32 Gold Star

										Touring	Scrambles	Clubman's and racing
Size	...	...	...	...	...	...	...	...	...	All ½ in x 0.305 in		
No. of pitches	...	...	...	...	...	...	...			67	66	68

B34 Gold Star

										Touring	Scrambles	Clubman's and racing
Size	...	...	...	...	...	...	...	...	...	All ½ in x 0.305 in		
No. of pitches	...	...	...	...	...	...	...			68	67	69

Gearbox oil capacity

All models	...	...	...	...	...	...	...	...	1 Imp pint (568 cc)
Oil viscosity	...	...	...	...	...	...	...	...	SAE 30 (winter) or 50 (summer) or Multigrade 20W/50

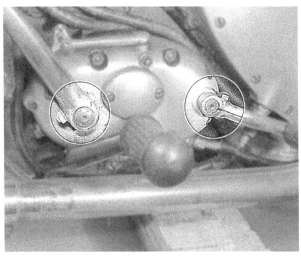

2.1a Kickstart lever is retained by cotter pin; gearchange lever by pinch bolt

2.1b Remove outer cover together with kickstart quadrant

1 General description

1 The range of models covered in this manual utilise two different types of four-speed gearbox, which are similar in principle but totally different in construction and layout. The gearbox fitted to all the M series models (also fitted to the pre 1955 B series machines) can be identified by the circular screw-type clutch adjustment cover and by the primary chain adjustment method, which is effected by a drawbolt fitted below the gearbox. The second type of gearbox, which is utilised by all B series machines and the Gold Start models, was introduced in late 1954 and has an improved method of gear selection and redesigned internal components. The gearbox shell is also substantially different. Variations of the later type gearbox were introduced especially for Gold Star models. The modifications include close-ratio gears and needle roller bearings in place of the plain bronze bushes. Available for use with rear-set footrest was a reverse camplate, which allowed the gearchange pedal to be reversed, but at the same time retaining the standard method of gear selection.

2 Either type of gearbox can be dismantled partially whilst it is still in the frame, since full access is gained by removing the inner and outer covers. The gearbox must be removed from the frame, however, if attention is required to the main bearing, final drive sprocket or the oil seal. In such cases, the gearbox can be removed without having to remove the engine, after detaching the gearbox mounting plates. As with gearbox dismantling in situ, the clutch and primary chaincase must be removed.

3 For ease of reference, the M series gearbox will be referred to as such and the later gearbox as B series.

2 Dismantling the gearbox: removing the outer and inner covers

B series only

1 Remove the drain plug from the base of the gearbox and drain off the oil. There is no need to remove either the kickstart or gearchange levers prior to removing the outer cover of the gearbox unless the kickstart return spring requires attention. It should be remembered, however, that the kickstart lever will be under tension whilst the outer cover is removed and will tend to unwind itself by causing the kickstart to rotate. This can be prevented by moving back the clutch operating lever at the top of the gearbox housing, so that it acts as a stop.

2 If it is desired to remove both the kickstart and the gearchange lever, the former is retained by a cotter pin that passes through the lower eye. Slacken the nut and tap the cotter pin

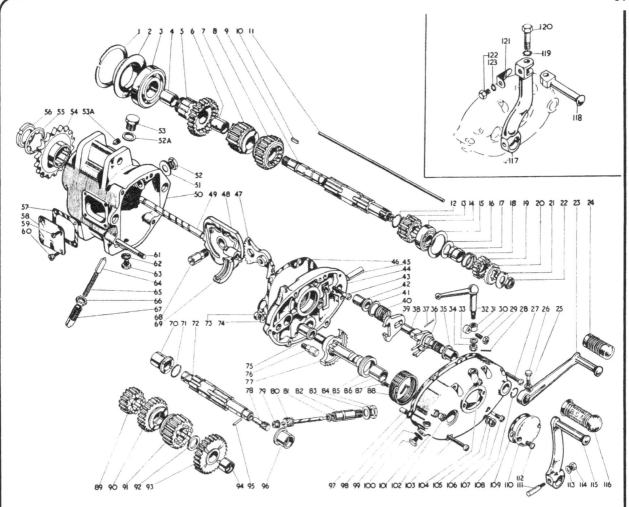

Fig. 2.1. B Series gearbox - component parts

1	Spring ring	33	Washer	63	Drain plug	95	Locating pin

1 Spring ring
2 Oil seal
3 Bearing
4 Bush
5 4th gear mainshaft pinion
6 Bush
7 2nd gear mainshaft pinion
8 3rd gear mainshaft piniont
9 Mainshaft
10 Woodruff key
11 Clutch pushrod
12 Spring clitp
13 1st gear mainshaft pinion
14 Mainshaft bearing
15 Spring ring
16 Washer
17 Bush
18 Ratchet spring
19 Ratchet pinion
20 Kickstart ratchet
21 Tab washer
22 Mainshaft nut
23 Gear change lever
24 Rubber
25 Washer
26 Screw
27 Screw - 2 off
28 Split pin
29 Nut
30 Adjusting screw
31 Pushrod lever
32 Clutch operating lever

33 Washer
34 Nut
35 Bush
36 Return spring
37 Split pin
38 Gear change spindle
39 Selector plate
40 Spring
41 Washer
42 Bush
43 Screw
44 Inner gearbox cover
45 Quadrant pivot pin
46 Paper gasket
47 Quadrant
48 Cam plate
49 Fork shaft
50 Gearbox casing
51 Washer
52 Nut
52a Washer
53 Plug
53a Grub screw
54 Gearbox sprocket
55 Lock washer
56 Nut
57 Inspection cover gasket
58 Inspection cover
59 Inspection cover screw
60 Inspection cover screw - 3 off
61 Stud - 4 off
62 Fibre washer

63 Drain plug
64 Camplate plunger
65 Camplate plunger spring
66 Locknut
67 Plunger housing
68 Pivot pin
69 Selector fork - 2 off
70 Layshaft bush
71 Locating ring
72 Layshaft
73 Nut
74 Washer
75 Bush
76 Kickstart quadrant stop
77 Kickstart spindle
78 Speedometer drive gear
79 Bush
80 Speedometer driven gear
81 Thrust washer
82 Bush
83 Fibre washer
84 Nut
85 Sleeve
86 Bush
87 Screw
88 Kickstart return spring
89 4th gear layshaft pinion
90 2nd gear layshaft pinion
91 3rd gear layshaft pinion
92 Thrust washer
93 1st gear layshaft pinion
94 Bush

95 Locating pin
96 Bush
97 Dowel - 2 off
98 Gasket
99 Fibre washer
100 Oil level plug
101 Outer cover
102 Locknut
103 Screw
104 Grease nipple
105 Washer - 4 off
106 Nut - 4 off
107 Speedometer driven gear retaining screw
108 Circlip
109 Fibre gasket
110 Inspection cover
111 Cotter pin
112 Screw - 2 off
113 Washer
114 Nut
115 Kickstart pedal
116 Rubber
117 Kickstart crank (folding type)
118 Kickstart pedal (folding type)
119 Washer
120 Bolt
121 Spring
122 Bolt
123 Washer

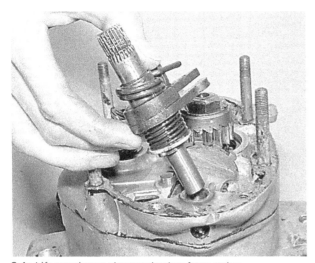

2.4a Lift gearchange claw mechanism from casing

2.4b Remove split pin to dismantle assembly

out of position to free its taper fit. It is advisable to retain the nut on the last few threads whilst the cotter pin is driven out, to prevent damage to them. The gearchange lever is mounted on splines and is retained by a pinch bolt arrangement. If the pinch bolt is slackened, the lever can be withdrawn from the splined shaft.

3   The end cover is retained by four nuts and three screws around the rim. When these have been slackened and removed, the end cover can be pulled away, exposing the gearchange mechanism.

4   The gearchange claw and return spring assembly will come away with the end cover and need not be dismantled unless replacements are necessary. The assembly is retained to the cover by the gearchange lever (if still fitted) and by a circlip. If the split pin is withdrawn from the spindle, the complete assembly can be stripped down.

5   Before the inner cover can be removed, it is necessary first to detach the kickstart ratchet pinion **held** to the end of the gearbox mainshaft by a nut and tab washer. Commence by withdrawing the clutch pushrod, then unscrew the nut, which has a normal right-hand thread, after bending back the tab washer. The two-part ratchet pinion will then pull off the splined end of the mainshaft, together with the ratchet spring bush and thrust washer. If the retaining screw of the inner cover is removed, the cover can be pulled away from the gearbox shell complete with the gearchange rocking lever, leaving the gear clusters in position.

6   Although it is unlikely that the gearchange rocking lever will need to be removed from the inner cover, it can be withdrawn by pushing the gear lever spindle bush out of the inner cover housing. This will reveal the end of the rocking lever spindle that has an internal ¼'' CEI thread. A screw or bolt of matching thread can be screwed into the spindle and used as a means of extraction.

3   Dismantling the gearbox: removing the gear clusters and selector arms

B series only

1   The rod on which the gear selector arms slide is pressed into the gearbox shell from the clutch end and is retained by a small grub screw on the outside of the case. Withdraw the grub screw and pull out the rod. It is now possible to withdraw the gear clusters and selector arms together with the layshaft, leaving only the final drive sleeve pinion and the cam plate within the gearbox shell.

2.5a Knock down tab washer ear to allow ...

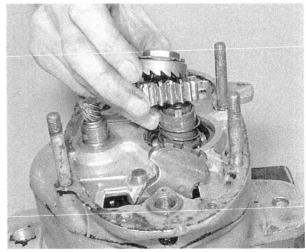

2.5b ... removal of ratchet nut and kickstart ratchet parts

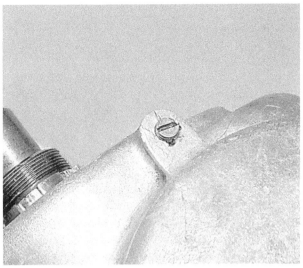
3.1a Unscrew retaining grubscrew to allow ...

3.1b ... removal of the selector fork rod

3.1c Loose 1st gear pinion may be removed

3.1d Lift mainshaft and selector fork out as unit ...

3.1e The layshaft complete with selector fork is then free

3.2a Cam plate will slide from pivot spindle after ...

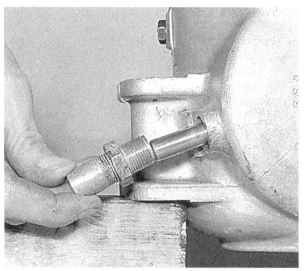

3.2b ... removal of detent plunger and housing

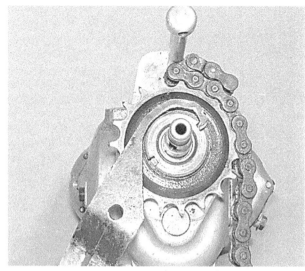

3.4a Sprag sprocket to loosen nut and allow ...

3.4b ... removal of nut, tab washer and sprocket

3.5a Drift sleeve pinion from the bearing

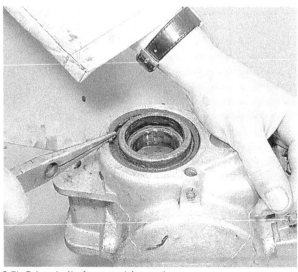

3.5b Prise circlip from position and ...

3.5c ... displace the oil seal holder plate

2 Unscrew the gear selector plunger housing locknut from the underside of the gearbox shell and remove the plunger housing complete. The housing contains a spring and the plunger that engages with the notches around the edge of the camplate. When the plunger assembly has been withdrawn, the camplate can be pulled from its pivot. If desired, the pivot too can be removed by bending back the external tab washer and unscrewing the retaining nut. It will be necessary to warm the gearbox shell in order to aid withdrawal.
3 The layshaft bearings are a press fit in the gearbox and can be driven out with a copper or brass drift. The mainshaft bearing within the inner cover can be knocked out of position by removing the retaining circlip and warming the cover.
4 To remove the final drive sprocket, wrap the final drive chain around the teeth and secure both ends of the chain in a vice. Bend back the tab washer and unscrew the large diameter retaining nut, which has a right-hand thread. The sprocket will pull off the splines of the sleeve gear pinion, which can then be tapped into the gearbox shell, through the main bearing, using a soft faced mallet.
5 The sleeve gear pinion should not be disturbed unless it has to be renewed. Before the bearing can be removed, it is necessary first to displace the oil seal. The oil seal is retained by a circlip, which must be prised out of position to release the seal. Warm the gearbox shell before the journal ball bearing is pressed out.

4 Dismantling the gearbox: removing the outer and inner covers

M series only
1 Remove the drainplug from the base of the gearbox and drain the oil. Remove the rectangular inspection plate from the top of the gearbox. Slacken the gearlever pinch bolt and withdraw the lever from the splined shaft. The kickstart lever need be removed only if attention to the kickstart quadrant or return spring is required. To remove the lever tap out the cotter pin after loosening the nut. Leave the nut in position, unscrewing it progressively as the cotter is drifted out. This will prevent damage to the threads.
2 Prise the circlip from the gearchange splined shaft. The gearbox outer cover is retained by seven cheesehead screws and four sleeve nuts, which once removed will allow removal of the outer cover together with the kickstart quadrant and return spring.
3 Disconnect the gearchange link rod at the forward end by detaching the split pin and withdrawing the clevis pin. Knock down the tab washer securing the nut on the end of the quadrant spindle, remove the nut and detach the link rod together with the operating arm.

3.5d Remove bearing only after heating case

4 The gearchange mechanism can be removed from the gearbox as a complete unit, further dismantling being necessary only if attention to the individual components is required. To remove the mechanism loosen and remove the large nut and washer from the exposed rear of the inner gearbox cover. Pull the complete assembly from the casing. Note the detent plunger unit, which is exposed when the gearchange mechanism is removed. Place the plunger and spring in a safe place to prevent loss. The mechanism can be dismantled after prising off the two pawl springs, using the blade of a screwdriver.
5 Place the gearbox, selector end upwards, in the soft jaws of a vice. Bend down the tab washer holding the outer nut on the mainshaft end. Loosen the nut and remove it together with the tab washer. Loosen and remove the inner nut which will free the kickstart ratchet assembly and tension spring. These components can be drawn off the splined mainshaft. Remove the gearbox from the vice.
6 The speedometer drive shaft and housing must be removed before the inner cover can be detached. Slacken the nut on the speedometer drive housing and give the drive end a shaft tap with a hammer, to loosen the housing. Remove the retaining grub screw from the housing tunnel in the inner cover. The drive housing is a tight push fit and will require drawing out of position. This may be done using the housing nut and spacers fitted progressively below the nut as the housing is drawn out. Lift the drive shaft from place.
7 The gearbox inner cover can now be detached after removal of the three retaining screws. Two of the screws have hexagonal heads and are secured by a shared locking plate, the edges of which must be bent down to free the screws. Tap the inner cover from position, ensuring that the selector quadrant comes away in situ. The selector quadrant need not be disturbed unless attention to it is required. Where necessary, push the quadrant from place and remove the detent plunger and spring from the inside of the inner cover.
8 As the inner cover is removed, the two gearshafts and the selector arms and rod will remain in position. Note the thick spacer washer and the thin shim on the end of the mainshaft.

5 Dismantling the gearbox: removing the gear clusters and selector arms

M series only
1 The complete gearbox shafts and gearclusters (except the sleeve pinion) together with the selector arm (fork) shaft and arms may be removed from the gearbox casing as a unit, to be further separated on the workbench.
2 To remove the final drive sprocket, wrap the final drive chain around the teeth and secure the ends of the chain in a vice. Bend back the tab washer and unscrew the slotted retaining nut, using a 'C' spanner or brass drift. The sprocket will pull off the splines of the top gear sleeve pinion, which can then be tapped into the inside of the gearbox, using a mallet.
3 The drive side ball bearing may be drifted from position, driving the oil seal with it, after heating the case in boiling water. The right-hand mainshaft bearing may be removed in a similar manner, after removal of the retaining circlip. The bronze layshaft bushes are a tight press fit and may be drifted from position, using a soft brass punch of 15/16 inch diameter.

6 Examination and renovation - general

1 Each of the various gearbox components should be examined carefully for signs of wear or damage after they have been cleaned thoroughly with a petrol/paraffin mix. A cleansing compound, such as "Gunk" or "Jizer" is particularly useful if the gearbox castings are covered with a film of oil and grease. Make sure all the internal parts have been removed if the gearbox has been dismantled prior to treatment, otherwise the subsequent water wash will cause rusting and damage to the bearings.

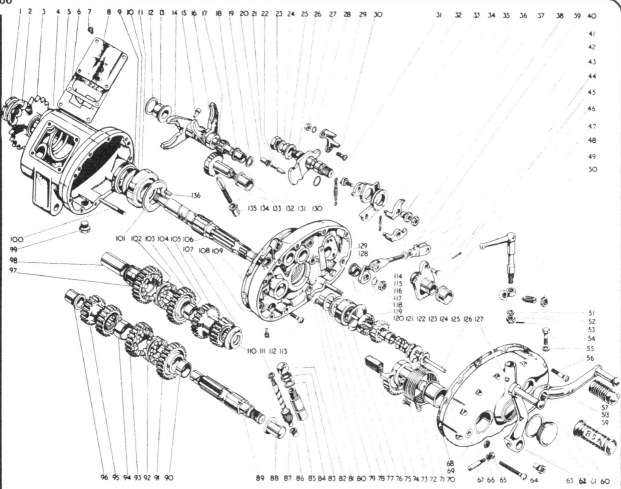

Fig. 2.2. M Series gearbox - component parts

1	Locknut	38	Locknut - 2 off	74	Kickstart quadrant	105	Mainshaft 2nd and 1st gear double pinion
2	Lockwasher	39	Bottom pawl	75	Rubber stop	106	Mainshaft
3	Gearbox sprocket	40	Link rod	76	Spring ring	107	Oil thrower
4	Gearbox casing	41	Locknut	77	Ball bearing	108	Inner cover
5	Baffle plate	42	Adjuster fork	78	Retainer	109	Gasket
6	Gasket - 2 off	43	Ratchet lever	79	Distance washer	110	Shim
7	Screw - 4 off	44	Clevis pin - 2 off	80	Stud	111	Thrust washer
8	Inspection cover	45	Split pin - 2 off	81	Dowel - 2 off	112	Grub screw
9	Oil seal q	46	Bush	82	Domed blanking nut	113	Screw - 3 off
10	Retainer	47	Clutch operating arm		(not fitted on speedo-	114	Operating lever
11	Bearing	48	Clutch pushrod lever		meter models)	115	Nut
12	Shim	49	Adjuster screw	83	Nut	116	Spring washer
13	Bush	50	Locknut	84	Washer	117	Spacer
14	Selector fork	51	Washer	85	Driven gear bush	118	Stud
15	Locating peg - 2 off	52	Castellated nut	86	Driven gear bush (small)	119	Kickstart ratchet
16	Selector fork	53	Split pin	87	driven gear	120	Kickstart ratchet pinion
17	Selector shaft	54	Bolt	88	Layshaft bush	121	Kickstart ratchet pinion sleeve
18	Bush	55	Washer	89	Layshaft	122	kickstart ratchet spring
19	Shim	56	Screw - 3 off	90	Bush	123	Kickstart ratchet pinion nut
20	Indexing plunger sleeve	57	Pedal rubber	91	Layshaft 1st gear pinion	124	Washer
21	Indexing plunger spring	58	Gear change pedal	92	Sliding dog	125	Locknut
22	Indexing plunger	59	Kickstart rubber	93	Layshaft 2nd gear pinion	126	Clutch pushrod
23	Nut	60	Fibre washer	94	Bush	127	Grease nipple
24	Washer	61	Inspection cap	95	Layshaft 3rd and 4th gear double pinion	128	Screw - 2 off
25	Bush	62	Kickstart pedal	96	Layshaft bush	129	Locking strip
26	Pedal shaft and plate	63	Nut - 4 off	97	Mainshaft 4th gear sleeve pinion	130	Circlip
27	Nut - 2 off	64	Screw - 4 off	98	Mainshaft 4th gear sleeve pinion bush	131	Quadrant
28	Washer - 2 off	65	Nut	99	Drain plug	132	Bush
29	Pawl arm stop	66	Washer	100	Stud - 2 off	133	Plunger
30	Screw - 2 off	67	Cotter pin	101	Oil retaining washer	134	Spring
31	Pawl arm spring	68	Gasket	102	Sliding dog	135	Cup
32	Pawl pivot - 2 off	69	Outer cover	103	Mainshaft 3rd gear pinion	136	Woodruff key
33	Pawl arm	70	Nut	104	Bush		
34	Pawl arm	71	Bush				
35	Pawl return spring	72	Kickstart return spring				
36	Top pawl	73	Anchor screw				
37	Nut - 2 off						

2 All gaskets and oil seals should be renewed, regardless of their condition, if the rebuilt gearbox is to remain oil tight. A rag soaked in methylated spirit provides one of the best means of removing old gasket cement, without having to resort to scraping and risk of damaging the mating surfaces.

3 Check for any stripped studs or bolt holes, which must be reclaimed before reassembly. Internal threads in castings can often be repaired cheaply by the use of what is known as a "Helicoil" thread insert, without need to tap oversize. Many motor cycle repairers can offer a "Helicoil" service.

7 Examination and renovation: gear pinions, selector arms and bearings

1 Examine each of the gear pinions for chipped, worn or broken teeth. Reject any that have such defects and fit new components. Check for worn dogs on the ends of each pinion, especially those where the edges are rounded. Worn dogs are a quite frequent cause of jumping out of gear and renewal of the pinions concerned is the only effective remedy.

2 Gear selection faults are often caused by bent selector arms or excessive side play in the arms as the result of seizure. Damage of this nature will be self-evident and replacement with new parts is essential.

3 Worn bearings are easy to renew, particularly those of the ball journal type. If a bearing of the ball journal type is suspect, it should be removed and washed out with a petrol/paraffin mix. If any signs of play are evident, or if the bearing runs roughly when it is turned, it should be rejected and renewed.

8 Examination and renovation: kickstart quadrant, ratchet and return spring

1 Examination of the kickstart quadrant will show the amount of wear that has taken place because it will be concentrated on the teeth that engage first with the ratchet pinion. Note that the leading tooth is relieved as standard, to prevent the quadrant from jamming during the initial engagement.

2 If the kickstart quadrant is worn, it is probable that the teeth of the ratchet pinion with which it engages will be worn too. Check the condition of the ratchet teeth. If the kickstart slips, it will be due to the rounding of the leading edge of these teeth. The complete ratchet assembly should be renewed when evidence of such damage is found, and the kick-

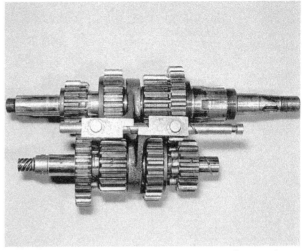
7.1 Gear clusters and selector forks - general view (B Series)

7.2a Pinion bush may be renewed

7.2b Check selector dogs and teeth for wear

8.1 Typical ratchet pawl wear causing jamming

8.4 Kickstart spring is held by screw

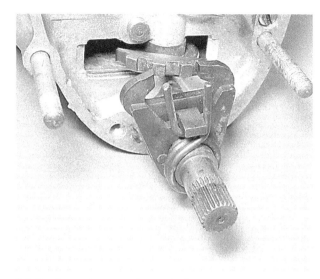

9.2 Check wear between claw and selector

9.3 Claw return spring must be fitted as shown

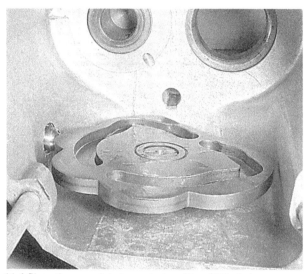

11.1 Place the cam plate in neutral position (B Series)

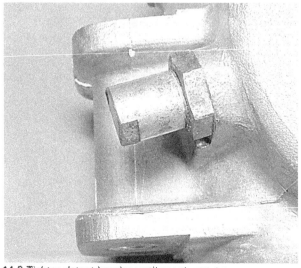

11.2 Tighten detent housing until one thread shows

11.3a Drive sleeve pinion into bearing

start ratchet if there is wear on the pinion teeth.
3   The light spring behind the ratchet assembly is unlikely to give trouble and need be renewed, only if it breaks or takes a permanent set.
4   It is good policy to renew the kickstart return spring during every major overhaul, whilst it is easy to fit. The spring leads a hard life, particularly if it is over-tensioned, and if it breaks, the kickstart must be tied up every time the engine is started, to bring the ratchet out of engagement.
5   The rubber kickstart quadrant stop fitted to the M series gearbox will become compacted or may disintegrate after a considerable time. The stop must be renewed to prevent damage to the aluminium outer cover.

9 Examination and renovation: gearchange mechanism and return spring

1   Wear is most likely to occur across the jaws of the gearchange claw. If the ends no longer have a sharp profile, the claw should be renewed.
2   The rocking arm quadrant with which the claw engages should also be renewed if the teeth show signs of wear. It is unlikely that one will show signs of wear without the other, unless a replacement has been made on some earlier occasion.
3   The gear selector camplate should be examined for worn cam grooves. Wear is unlikely unless the machine has covered a considerable mileage or if the gearbox has been run for a lengthy period with low oil content.
4   Check the tension of the gear selector plunger spring and replace the spring if it no longer maintains good pressure on the plunger that engages with the edge of the gearchange camplate. Imprecise gearchanges, or jumping out of gear can sometimes be attributed to this type of fault. Correct tension is normally achieved by having about one thread of the adjuster showing. after the locknut has been tightened.

10 Examination and renovation: gearchange mechanism

M series only
1   Inspect the ratchet plate teeth and the pawls with which they engage for wear. Pronounced wear will increase the gear lever movement required when changing gear and in extreme cases may prevent selection.
2   Renew the pawl springs if there are signs of stretching and the resultant loss of tension. Check the spring ends for wear, which may indicate imminent breakage.
3   Check the tension of the gear selector detent plunger spring and the selector quadrant detent spring. Replace either spring if tension is suspect.
4   Test the play in the selector link rod clevis pins and forks. Wear will produce indefinite gear selection, and may when combined with worn ratchet plate teeth and pawls, create total loss of selection.

11 Reassembling the gearbox: replacing the camplate and gear clusters

B series only
1   Assuming the main bearing and oil seal (if fitted) in the gearbox shell have been left in position or replaced, the camplate can be fitted in conjunction with the camplate plunger assembly. Fit the camplate so that the plunger engages with the neutral cutaway; this is much shallower than the cutaways for the individual gear selections and is located mid-way between the cutaways for first and second gear. It is essential that the gearbox is assembled with the selectors in the neutral position, otherwise gear indexing will not be correct.
2   The camplate plunger should be adjusted so that engagement is effected in a positive manner without the operating action

becoming too stiff. Adjustment is correct if approximately one thread is showing, when the adjuster locknut is tightened.
3   Fit the mainshaft sleeve gear making sure it extends fully through the gearbox main bearing. Replace the layshaft complete with the three innermost gears, after lubricating the bearing bush in the gearbox shell. The lower selector arm should engage with the raised tracks where the two middle gear pinions abut one another; it is best to have the selector arm in position when the layshaft and gear pinions are inserted into the gearbox shell. Check that the selector arm has remained engaged with both pinions and that it has engaged with the lower track of the camplate. Fit the two middle mainshaft pinions with the upper selector arm, which must engage with their raised tracks and the upper track of the camplate in a manner similar to that of the lower arm. Slide the gearbox mainshaft through both pinions and the sleeve gear, then fit the selector rod with the cutaway in the end nearest to the gearbox main bearing. Anchor the selector rod with the grub screw that passes through the gearbox shell and stake the end of the grub screw hole so that it cannot work out and fall free.
4   Assembly of the gearbox internals is often easier if the side cover is removed. The cover takes the form of a square plate in

11.3b Refit the layshaft and selector fork followed by ...

11.3c ... two middle mainshaft pinions and selector fork

11.3d Insert mainshaft and ...

11.3e ... secure selector forks with rod

11.4a Apply locking fluid to the grub screw

11.4b Replace the 1st gear pinion

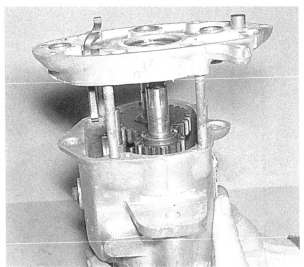

12.1a Position the gearbox inner cover and ...

12.1b ... align the timing marks before fitting

the rear of the gearbox shell that is held by four screws and seats on a paper gasket. When it is removed, it is easier to check whether the gear selector arms are located correctly with both the pinions and the camplate.

5 Replace the two end pinions on the layshaft and mainshaft and re-check whether the gearchange is still in the neutral position.

12 Reassembling the gearbox: refitting the inner cover and kickstart ratchet

B series only

1 Replace the gearbox inner cover, which is retained in position by a single, cheese head screw. Use a new paper gasket at the joint, which should be lightly coated with gasket cement. Before the end cover is pushed home fully and tightened down, it is imperative to ensure that the red dot on the rocking arm quadrant is in line with the red dot on the cover and is maintained in this position until the cover is fitted. FAILURE TO OBSERVE THIS PROCEDURE WILL RESULT IN INCORRECT INDEXING OF THE GEARS. On some gearboxes small dimples take the place of the red dots.

2 Check that the gearbox is still in the neutral position, then replace the square plate in the rear of the gearbox shell. A new jointing gasket should be fitted and gasket cement applied to ensure an oiltight joint. Make sure each of the screws has a spring washer, to prevent slackening.

3 Replace the metal bush on which the kickstart ratchet pinions seats, the light coil spring, then the kickstart ratchet pinion and the other half of the ratchet assembly. Refit the shouldered nut and tab washer that locks the whole assembly in place and tighten before bending the tab washer into the locking position.

4 Check that the gearbox mainshaft and sleeve gear still revolve freely.

13 Reassembling the gearbox: refitting the outer cover

B series only

1 The end cover should contain both the kickstart and kickstart quadrant assembly and the gear lever and gear change operating claw. If either or both assemblies have been removed, they must be replaced by reversing the dismantling procedure.

2 Tension the kickstart return spring by turning the kickstart anti-clockwise and prevent it from unwinding by moving back the clutch operating lever so that it acts as a temporary stop.

12.3a Refit the backing washer and bush followed by ...

12.3b ... the ratchet spring and pinion and ...

12.3c ... the ratchet pawl, nut and tab washers

13.2a Use clutch arm to hold kickstart lever when ...

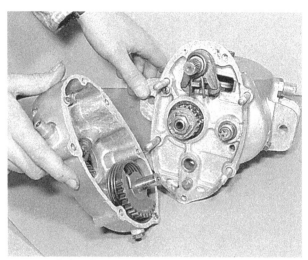

13.2b ... refitting the outer cover

13.3a Assemble gearbox plates loosely before ...

13.3b Positioning gearbox in frame

Apply gasket cement to the outer face of the inner cover and to the inner face of the outer cover, using a new gasket at the joint face. Push the outer case on to the four studs that project from the inner cover and depress the kickstart slightly, so that the quadrant will clear its stop. Push the cover home fully, then replace the four nuts and three screws; tightening them fully. Check that the kickstart will turn the mainshaft as it is depressed and that the gearchange lever will select all four gears in the correct sequence. It may be necessary to revolve the mainshaft and sleeve gear by hand at this stage, so that the sliding dogs will engage correctly.

3   If all is correct the gearbox can now be replaced in the frame and the primary drive components replaced.

4   Refill the gearbox with clean engine oil. All gearboxes have an oil content of 1 pint (570 cc). Continue adding oil until it commences to emerge from the level plug and replace the level plug after all excess oil has drained off.

5   Replace the oval end plate in the centre of the end cover, after checking to ensure the clutch push rod is in position. This cover has a cork gasket and is retained by two screws. No gasket cement is needed.

14 Reassembling the gearbox: replacing the gearclusters and inner cover

M series only

1   Refit the drive side mainshaft bearing after warming the case. Fit the smaller of the two washers between the bearing and the case, and the larger washer between the bearing and the top gear sleeve pinion. Carefully drift the sleeve pinion into position after replacing the oil seal. The lip of the seal should be lubricated to accept the pinion sleeve.

2   The two gearshafts, complete with gear pinions and the selector fork assembly, should be assembled as a unit and introduced into the gearbox placed in the top gear position. To find top gear position rotate the selector fork spindle until the mainshaft selector dog is engaged with the top gear pinion.

3   Insert the complete assembly into the gearbox shell and check that the gears are still in the top gear position by viewing through the inspection window.

4   Place the selector quadrant into position in the inner cover, together with the detent plunger and spring. Rotate the quadrant until it is in the top gear notch. (see diagram).

5   Place a new gasket in position on the gearbox shell after liberally coating both sides of the gasket with jointing compound. Replace the large washer followed by the small washer onto the end of the mainshaft. Position and replace the inner cover. If necessary, turn the selector quadrant shaft fractionally, with a spanner, to aid meshing of the quadrant and the selector spindle teeth. Insert and tighten the three inner cover retaining screws. Do not omit the locking plate, which secures the two front screws.

6   Before continuing, rotate the selector quadrant shaft by applying a spanner to the flats, and check that each gear can be selected.

15 Reassembling the gearbox: replacing the gearchange mechanism and outer cover

M series only

1   Assemble the gearchange mechanism as a complete unit, referring to the line drawing to aid correct positioning of components. Fit the mechanism into the casing and replace and tighten the nut and washer. Do not omit the small detent plunger and spring.

2   Refit the selector link rod to the quadrant shaft and reconnect the rod with the gearchange mechanism. Do not omit the split pin which secures the clevis pin. Temporarily refit the gearchange lever and check that each gear can be selected. If the selector dogs do not appear to go right home when in top or bottom gear, loosen the locknut on the link rod, remove the clevis pin

and alter the length of the rod a little at a time until the mal-adjustment is corrected.

3   Fit the kickstart ratchet assembly commencing with the bush followed by the ratchet, the ratchet pinion and the spring and sleeve nut. The sleeve nut should only be tightened finger tight. Fit the tab washer to the mainshaft so that the tongue engages with the splined shaft end. Fit and tighten the locking nut fully and secure it by bending up the tab washer.

4   Place the speedometer drive shaft into the housing sleeve and gently drift the sleeve into the housing tunnel in the inner cover. Take care that the shaft gear engages correctly with the gear on the layshaft. Fit and tighten the housing sleeve grub screw. Replace the washer and nut on the housing thread.

5   The outer cover may now be replaced after placing a new gasket in position. Once again apply gasket compound to ensure an oiltight joint. Fit and tighten the seven screws and four nuts which secure the cover.

6   The gearbox is now ready for replacement in the frame followed by refitting of the gearbox sprocket and primary drive components.

7   Refill the gearbox with 1 pint (570 cc) of engine oil after replacing the inspection cover and drain plug. Continue filling until oil starts to emerge from the level orifice. Fit and tighten the level plug.

Fig. 2.3. Selector quadrant 'M' models only

16 Fitting a reverse camplate

Gold Star models only

1   When rearset footrests are fitted, it is advantageous to reverse the gearchange lever, so that it faces backward and is closer to the repositioned footrests. Unfortunately, this also reverses the gear positions, so that the pedal must be depressed to select bottom gear and raised to select second, third and top gears. There is risk of unwittingly selecting the wrong gear due to familiarity with the original arrangement, which may cause the engine to over-rev and sustain mechanical damage. In consequence a reverse camplate is available as an optional extra, to ensure the original sequence of gear operation can be retained with the reverse pedal.

2   The gearbox must be stripped completely to fit the reverse camplate, by following the procedure detailed in Section 2 to 3 of this Chapter. Note that when the gearbox is reassembled, the red dot marks on the rocking arm quadrant and inner cover will no longer coincide for correct meshing. Fitting is by trial and error.

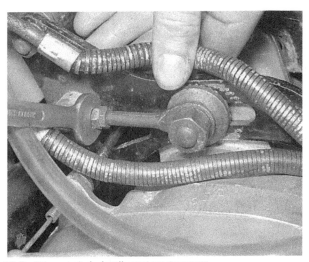

18.2a Primary chain is adjusted by drawbolt

17 Changing sprocket sizes

1   Apart from the need to fit a larger rear wheel sprocket or smaller engine sprocket in order to lower the overall gear ratios when a sidecar is attached, it is seldom that any real advantage is gained from departing from the manufacturer's recommended sizes. Many owners believe that raising the overall gear ratios will give an immediate increase in maximum speed, without any further attention to the engine. This is by no means correct since it is only too easy to convert top gear into an 'overdrive', with the result that the machine will be as fast or even faster in third gear under certain conditions.

2   A tuned engine may benefit from raising the overall gear ratios because the tuning operation itself will tend to concentrate the power band higher up the scale and raise the peak rpm at which maximum power is available. Such changes must invariably be made on a trial and error basis, since there is no reliable means of predicting how a newly tuned engine will perform in these circumstances. It is wise to make initial tests with the standard ratios unchanged, so that they can be used as the basis with which to compare all further changes in gear ratios. It is fatal to change more than one variable at a time, otherwise there is no means of assessing to which factor an improvement can be attributed.

18.2b Gearbox pivots on lower bolt; upper bolt slides

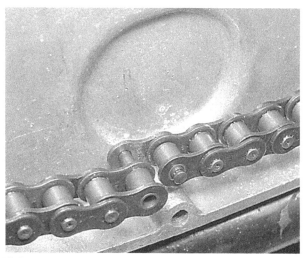

18.4 Depression in case aids primary chain removal

18 Primary chain: examination, lubrication and adjustment

1 Although the primary chain runs under ideal conditions, in the sense that it is totally enclosed and is running through an oil bath, it will nonetheless require periodic attention to take up any slackness that has developed as the result of wear.
2 Adjustment of the chain should be made by moving the gearbox forward or backwards, after slackening the mounting bolts, by means of a drawbolt fitted to the right-hand side of the gearbox. The chain tension is correct when ½ inch up and down play can be felt in the centre of the upper chain run; access can be made through the inspection cap/oil filler orifice.
3 Always check the tension with the chain at the tightest point. Chains and sprockets rarely wear evenly and may give one or more tight points at a certain position in rotation.
4 All models are fitted with a single row chain that has a spring connecting link. A small depression formed in the lower rear half of the primary chaincase facilitates removal and replacement of this link. Replacement of the chain is required when the amount of play lengthwise exceeds ¼ in per foot.
5 If the machine is used at irregular intervals, or for a series of short journeys, the oil in the primary chaincase should be changed frequently, to offset the effects of condensation. A rusty brown deposit on the chain links is the first obvious sign that condensation is contaminating the oil.

19 Speedometer drive

1 Unlike many other British motor cycles, the speedometer drive is taken from the end of the gearbox layshaft and not the rear hub. A worm attached to the end of the layshaft transmits the drive through a similar worm mounted at right-angles, so that the drive cable can be attached to the forward facing end of the gearbox.
2 Because the speedometer drive pinions work in well-lubricated conditions, they are unlikely to give trouble during the normal service lift of the machine. Both can be replaced in the event of drive failure; it is always advisable to renew the drive pinions as a matched pair.

20 Fault diagnosis: gearbox

Symptom	Cause	Remedy
Kickstart does not return when engine is turned over or started	Broken kickstart return spring	Renew.
Kickstart slips and will not turn engine over	Worn ratchet assembly	Renew.
Kickstart jams	Worn ratchet assembly and/or kickstart quadrant	Renew.
Difficulty in engaging gears	Gear selectors not indexed correctly	Check alignment of 'timing' marks on inner cover (B series only).
	Selector pawls not engaging due to wear	Renew pawls (M series only).
	Weak gearchange mechanism springs	Renew springs (M series only).
	Selector forks bent or badly worn	Renew.
Machine jumps out of gear	Camplate plunger tensioned incorrectly	Tighten adjustment (B series). Renew spring (M series).
	Worn dogs on gear pinions	Renew defective pinions.
Gearchange lever does not return to original position	Broken return spring	Renew spring in outer cover (B series).
	Weak gearchange mechanism springs	Renew springs (M series).
Gearchange lever will not select any gears.	Camplate plunger adusted too tightly	Slacken off adjustment

Chapter 3 Clutch

Contents

General description 1
Dismantling the clutch 2
Clutch plate: examination and renovation 3
Clutch inner and outer drums: examination and
renovation 4
Clutch centre: examination and renovation 5
Clutch springs: examination and renovation 6
Reassembling the clutch... 7
Clutch adjustments: general 8
Fault diagnosis 9

Specifications

No. of plates

	Plain	Friction
M20, 21 and 33	4	5 *
B31 pre 1958	3	4 *
B33 pre 1958	4	5 *
B31 and 33 post 1957	5	5
B32, B33 Gold Star	5	6 *
B34 Gold Star	5	6 *

* Including the clutch outer drum/chain wheel

Thickness of linings **

Chainwheel	0.300 in - 0.310 in (7.62 - 7.87 mm)
All other linings	0.155 in - 0.165 in (3.93 - 4.19 mm)

** Does not apply to 1958 - 60 B31 - 33 models

1 General description

1 Two different types of clutch have been fitted to the BSA 350 - 600 cc range of pre-unit construction engines. The original design which was used on all models except the B series alternator models, comprises a number of plain and friction type plates sandwiched under pressure from six springs adjustable by means of nuts and locknuts. The clutch outer drum/chainwheel serves as an additional friction plate into which bonded fibre inserts are fitted. The clutch unit runs on a double row caged ball bearing. The B series alternator model clutch is similar in principle but does not utilise the chainwheel as a friction plate, and has only four pressure springs, each of which is adjustable by means of a special slotted sleeve nut. The clutch runs on twenty uncaged rollers.

2 Dismantling the clutch

1 Either type of clutch can be dismantled without need to remove the clutch centre from the gearbox mainshaft. After the chaincase oil has been drained and the outer cover removed, the clutch is dismantled by removing the clutch adjusting nuts, to release the tension springs. Where the special slotted nut is used a suitable screwdriver can be fabricated for removal and refitting and also for subsequent adjustment.
2 When the tension springs have been withdrawn, the pressure plate and the inserted and plain plates can be withdrawn. There is no necessity to detach the primary chain, unless the clutch

chainwheel requires attention.
3 If the clutch chainwheel has to be removed, it will be necessary to bend back the tab washer that locks the retaining nut of the inner drum. This has a right-hand thread. The inner drum will pull off the splines of the clutch centre, releasing the clutch chainwheels. The primary chain must be detached before the chainwheel will pull off the centre bearing.
4 It is possible that the rollers from the clutch centre bearing will fall free as chainwheel is removed, (alternator models only). It is best to use thick grease to retain the rollers in place, or there is danger of them being lost or misplaced during reassembly.

3 Clutch plates: examination and renovation

1 Check each clutch plate to make sure that it is completely flat and free from buckles. Reject any that have become distorted otherwise clutch troubles will persist.
2 The tongues at the edges of each inserted plate should be free from burrs or other damage. It is permissible to remove any burrs by dressing the edges of each tongue with a file. Make sure they remain square. Reject any plates where the tongues show signs of splitting or cracking. A repair is impracticable especially in view of the relatively cheap cost of replacements.
3 Check also the tongues at the centre of each plain plate. Make sure the surface of each plate is smooth and free from blemishes.
4 The linings of each inserted plate should be inspected. The amount of wear that has taken place can be checked against the Specifications at the beginning of this Chapter: if there is any doubt, renew the plates whilst the clutch is dismantled.

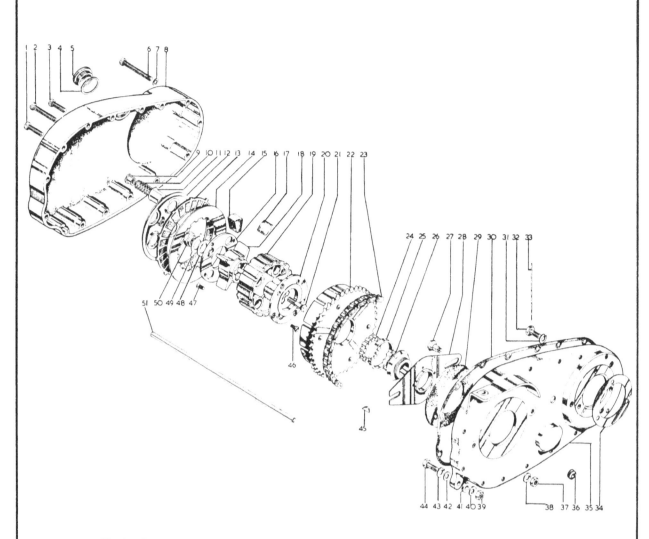

Fig. 3.1. Later type clutch and primary chaincase – component parts (1958 on alternator models)

1	Screw - 6 off	25	Clutch bearing
2	Level and drain screws - 2 off	26	Clutch sleeve
3	Screw - 4 off	27	Bolt - 2 off
4	Fibre washer	28	Sliding plate
5	Inspection cap	29	Felt washer
6	Screw - 3 off	30	Gasket
7	Fibre washer - 5 off	31	Spring washer - 3 off
8	Outer chaincase	32	Bolt
9	Spring tension nut - 4 off	33	Locking wire
10	Clutch spring - 4 off	34	Gasket
11	Clutch spring cap - 4 off	35	Inner chaincase
12	Clutch pressure plate	36	Grommet
13	Friction plate - 5 off	37	Nut - 2 off
14	Driven plate - 5 off	38	Washer - 2 off
15	Shock absorber rubber - 4 off	39	Nut
16	Outer retaining plate	40	Washer
17	Rebound rubber - 4 off	41	Spacer
18	Shock absorber spider	42	Washer
19	Clutch centre	43	Spring washer
20	Inner retaining plate	44	Bolt
21	Centre pin - 4 off	45	Woodruff key
22	Housing and chainwheel	46	Screw -4 off
23	Primary chain	47	Screw - 4 off
24	Clutch rollers - 20 off	48	Washer
		49	Tab washer
		50	Nut
		51	Clutch pushrod

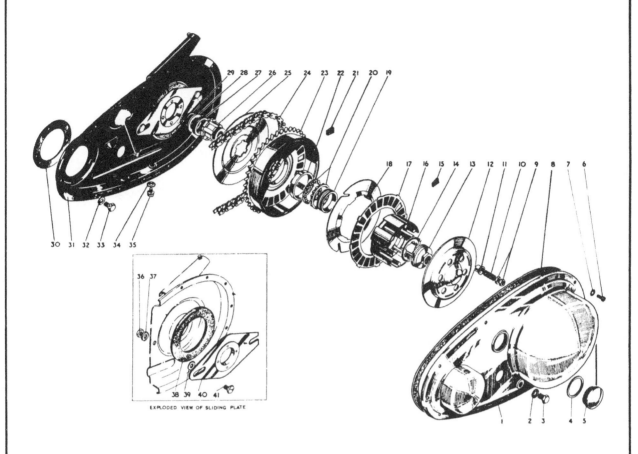

EXPLODED VIEW OF SLIDING PLATE

Fig. 3.2. Early type clutch and primary chaincase – component parts (rigid and plunger models)

1 Outer chaincase	12 Clutch cover	20 Ballrace retaining ring - 2 off	30 Hollite gasket
2 Fibre washer	13 Mainshaft nut	21 Chainwheel insert - 24 off	31 Inner chaincase
3 Oil level plug	14 Tab washer	22 Ballrace outer ring	32 Spring washer - 3 off
4 Fibre washer	15 Clutch plate insert - 72 off	23 Chainwheels	33 Bolt - 3 off
5 Inspection cap	16 Clutch centre	24 Primary chain	34 Fibre washer
6 Screw - 16 off	17 Clutch friction plate - 3 or 4 off	25 Clutch plate	35 Drain plug
7 Washer - 16 off	18 Clutch plain plate 1, 2 or 4 off	26 Driving sleeve	36 Nut - 2 off
8 Cork gasket	19 Ballrace inner ring	27 Sleeve	37 Washer - 2 off
9 Nut - 12 off		28 Retaining washer	38 Felt washer
10 Clutch spring - 6 off		29 Felt washer	39 Fibre washer - 2 off
11 Spring cap - 6 off			40 Sliding plate
			41 Bolt - 2 off

1   Oil level and drain screw – 2 off
2   Fibre washer – 2 off
3   Screw – 7 off
4   Nut – 12 off
5   Clutch spring – 6 off
6   Spring cap – 6 off
7   Clutch cover
8   Clutch plate inserts
9   Ballrace inner ring
10 Ballrace retaining ring – 2 off
11 Chainwheel insert – 24 off
12 Primary chain
13 Driving sleeve
14 Sleeve collar
15 Bolt – 2 off
16 Spring washer – 2 off
17 Nut – 2 off
18 Bolt(s)
19 Spacer
20 Spring washer
21 Nut
22 Dowel – 2 off
23 Inner chaincase
24 Felt backwasher
25 Sliding plate
26 Clutch plate
27 Chainwheel
28 Ballrace outer ring
29 Gasket
30 Clutch plain plate(s)
31 Clutch friction plate(s)
32 Clutch centre
33 Lock washer
34 Nut
35 Fibre washer

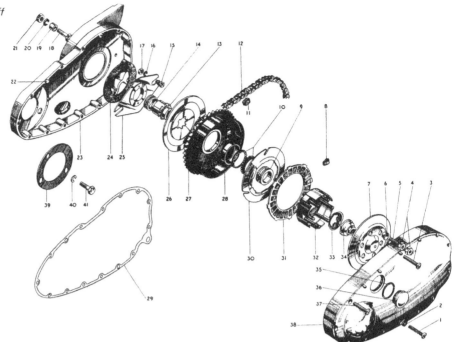

Fig. 3.3. Early type clutch and primary chaincase – component parts (1954 to 1958 – all swinging arm models, including Gold Star)

36 Inspection cap
37 Screw – 6 off
38 Outer chaincase

39 Washer
40 Spring washer – 3 off
41 Bolt – 3 off

4   Clutch inner and outer drums: examination and renovation

1   Check the condition of the slots in the inner and outer drums. After an extensive period of service, the tongues of the clutch plates tend to make indentations in the edge of each slot, which eventually prevent the clutch from freeing completely. The edges should be dressed with a file until they are square and completely free from indentations. Provided too much metal has not been removed, the additional amount of backlash in the plates will not be of too great consequence.

2   Inspect also the condition of the linings within the outer drum and the teeth of the integral chainwheel sprocket, (later models only). The linings of the outer drum assembly are thicker than those of the inserted clutch plates - see Specifications at the beginning of this Chapter. Check the condition of the outer race in the centre of the outer drum assembly. This must be smooth and not chipped or indented. A noisy clutch can often be traced to a badly-worn centre bearing.

3   Check that the studs used for the clutch tensioners are riveted firmly to the inner drum and are not bent. Check that the threads are in good condition. The innermost tongues of the inner drum must make good contact with the splines of the clutch centre. Backlash will impose a shear stress on the tongues, if wear is evident, leading to early failure.

5   Clutch centre: examination and renovation

1   There is no necessity to withdraw the clutch centre from the mainshaft unless the gearbox mainshaft requires removal or a loose fit has developed between the clutch centre and the tapered mainshaft.

2   Under normal circumstances the clutch centre is a very tight fit on the mainshaft and will require drawing from position using BSA service tool No. 61-3362 which screws into the centre. A two or three legged sprocket puller may also be utilised, although

it may be necessary to slacken off the primary chain case inner mounting screws to give sufficient clearance for the puller.

3   If backlash develops between the clutch centre and the centre boss, one or both components should be renewed. A loose fit between the centre and the mainshaft may be restored by grinding-in with fine valve grinding compound, after the key has been removed temporarily. Caution is necessary, however, otherwise the clutch centre will seat too far back and permit the clutch to foul the back of the rear half of the chain case.

4   Most cases of damage to the clutch centre and/or gearbox mainshaft can be attributed to the retaining nut working loose. It is imperative that this nut is tightened fully during reassembly, and locked securely with the tab washer provided.

5   It is also advisable to check the condition of the threads on the end of the gearbox mainshaft and on the retaining nut itself. Replacement parts must be fitted if the condition of either is questionable.

6   Clutch springs: examination and renovation

1   After an extensive period of service, the springs will begin to take a permanent set, necessitating their replacement. Stand the springs on end and compare them in length with a new spring. If they are shorter in length, replace the whole set.

2   Springs should always be replaced as a complete set, even if only one is shorter in length. This will aid subsequent adjustment and facilitate arriving at the correct individual tension.

7   Reassembling the clutch

1   Reassemble the clutch by reversing the procedure detailed in Section 2 of this Chapter. Make sure the clutch plates are alternated correctly and that a small dab of grease is applied to the centre of the pressure plate, at the point on which the end of the clutch pushrod bears.

2 Adjust the clutch spring tensioners so that the extreme end of the thread on the studs is just showing. Check the clutch action by means of the kickstart, and readjust if necessary, before fitting the outer chaincase (and domed cover, early models). Some adjusting nuts may require tightening more than others to compensate for minor variations in spring tensions.

8 Clutch adjustments: general

1 Apart from the clutch spring tensioners, there are two other means of adjusting the clutch action, mainly with regard to the operation of the handlebar-mounted lever. A cable adjuster is mounted on a lug at the top of the gearbox casing so that slack can be taken out of the clutch cable as the clutch beds down or as the cable stretches, and there is a screw and locknut within the outer cover of the gearbox (access via oval cover plate retained by two screws) that controls the action of the control operating arm.
2 The clutch cable is adjusted correctly if there is about 1/16 in (1.5 mm) free play at the handlebar lever before the clutch

commences to operate. It is important that this minimum amount of play is maintained at all times. FAILURE TO OBSERVE THIS RULE WILL IMPOSE A HEAVY LOAD ON THE CLUTCH PUSHROD, LEADING TO RAPID WEAR AND LOSS OF ADJUSTMENT. Many owners have been perplexed by the continual loss of clutch adjustment, caused by a rapidly shortening pushrod. Failure to maintain this small but vital amount of free play causes the pushrod to overheat under the continually-applied load, thereby softening the hardened ends and promoting advanced wear.

3 Heavy clutch action is sometimes caused by the poor operating angle of the external clutch operating arm on the top of the gearbox. To gain the advantage of maximum leverage, the operating lever should be adjusted so that it lies parallel to the gear box, WHEN THE CLUTCH IS FULLY DISENGAGED.
4 Adjustment of the clutch springs can be made only after the primary drive cover has been removed.
5 Clutch drag will result if the tension is too great or if the springs have not received individual attention to ensure the clutch plates separate squarely.

8.2a Adjust clutch at adjustment screw and ...

8.2b ... then adjust slack in cable

9 Fault diagnosis - clutch

Symptom	Cause	Remedy
Engine speed increases but not road speed	Clutch slip; incorrect adjustment or worn linings	Adjust, or renew clutch plates.
Machine creeps forward when in gear; difficulty in finding neutral	Clutch drag; incorrect adjustment or damaged clutch plates	Readjust, or fit new clutch plates.
Machine jerks on take-off or when changing gear	Clutch centre loose on gearbox mainshaft	Check for wear and retighten retaining nut.
Clutch noisy when withdrawn	Badly worn clutch centre bearing	Renew bearing.
Clutch neither frees or engages smoothly	Burrs on edges of clutch plates and slots in clutch drums	Dress damaged parts with file if damage not too great.
Clutch action heavy	Overtight tension springs or wrong angle of operating arm	Slacken tension nuts or readjust operating arm.
	Dry operating cable, or bends too tight	Lubricate cable and reroute as necessary.
Clutch action harsh	Overtight primary chain	Readjust primary chain.
Clutch 'bites' at extreme end of lever movement	Worn linings	Renew clutch plates (inserted).
Constant loss of clutch adjustment	Worn pushrod, due to failure to maintain minimum clearance.	Renew pushrod and readjust.

Chapter 4 Fuel system, carburation and lubrication

Contents

General description 1	Exhaust systems 9
Petrol tank: removal and replacement 2	Engine lubrication system 10
Petrol taps: removal and replacement 3	Lubrication: maintenance 11
Petrol feed pipes: examination 4	Oil filters: cleaning 12
Carburettor: removal 5	Removing, dismantling and reassembling the oil pump ... 13
Carburettor: dismantling, examination and reassembly ... 6	Fault diagnosis: fuel system, carburation and engine
Carburettor: checking the settings 7	lubrication 14
Air filter: location, examination and cleaning element ... 8	

Specifications

Petrol tank capacity

M series	3 galls/3.5 US galls (13½ litres)
B series except GS models	3 galls/3.5 US galls (13½ litres)
Gold Star models:	
Touring	4 galls/4.68 US galls (18 litres)
Scrambles	2 galls/2.34 US galls (9 litres)
Racing	4 galls/4.68 US galls (18 litres)
Clubmans	5 galls/5.90 US galls (22.7 litres)

Oil tank capacity

M series	5 Imp pints
B series except GS models	4 Imp pints
Gold Star models:	
Touring	5½ Imp pints
Scrambles	5½ Imp pints
Racing	5½ Imp pints
Clubmans	5½ Imp pints

Primary chaincase capacity

M series	56 cc (2 fl oz)
All B series except alternator models	225 cc (8 fl oz)
Alternator models	284 cc (10 fl oz)

Carburettor settings

Gold Star models, petrol or petrol-benzole fuel

Make: Amal   Type: Monobloc

Model	Application	Choke Diameter	Main Jet	Throttle Valve	Needle Position	Needle Jet
CB32GS	Scrambles	1 1/16 in	260	4	3	.1065
DB32GS	Scrambles	1 1/16 in	260	4	3	.1065
CB34GS	Scrambles	1 1/8 in	240	3	2	.1065
DB34GS	Scrambles	*1 5/32 in	320	3	2	.1065
DBD34GS	Scrambles	*1 5/32 in	320	3	2	.1065

Early 1955 scrambles models were fitted with 1 1/8 in choke, 240 jet monobloc.

Make: Amal   **Type: 10TT**

Model	Application	Choke Diameter	Main Jet	Throttle Valve	Needle Position	Needle Jet
CB32GS	Touring	1 3/32 in	360	7	3	.109
CB32GS	Scrambles	1 3/32 in	360	7	3	.109
CB32GS	Racing	1 3/32 in	360	7	3	.109
DB32GS	Touring	1 3/32 in	360	7	3	.109
CB34GS	Touring	1 3/16 in	360	7	3	.109
CB34GS	Racing	1 3/16 in	360	7	3	.109
DB34GS	Touring	1 3/16 in	360	7	3	.109
DBD34GS	Touring	1 3/16 in	360	7	3	.109

Make: Amal   **Type: Grand Prix (G.P.)**

Model	Application	Choke Diameter	Main Jet	Throttle Valve	Needle Position	Needle Jet
CB32GS	Racing	1 3/32 in	210	6	3	.109
DB32GS	Racing	1 3/16 in	280	5	4	.109
CB34GS	Racing	1 7/32 in	260	7	2	.109
DB34GS	1955 Racing	1 3/8 in	330	7	4	.109
DBD34GS	1955 Racing	1 3/8 in	330	7	4	.109
DB34GS	1956 Racing	1½ in	†350	4	3	.109
DBD34GS	1956 Racing	1½ in	†350	4	3	.109

† 390 main jet used for road racing with megaphone.

Carburettors: Gold Star models, petrol or petrol-benzole

Make: Amal   **Type: Remote needle (R.N.)**

Model	Application	Choke Diameter	Main Jet	Throttle Valve	Needle Position	Needle Jet
CB32GS	Racing	1 3/32	450	6	4	.109
CB34GS	Racing	1 3/16	520	7	4	.109

Carburettors: all models except Gold Stars, petrol

Make: Amal   **Type: 376 Monobloc**

Model	Choke Diameter	Main Jet	Pilot Jet	Throttle Valve	Needle Position	Needle Jet	Air Filter	Carb. Model
M.20	1 in	240		376/5	3	.1065	Yes	376/21
M.21	1 1/16 in	250	30	376/5	2	.1065	Yes	376/23
M.33	1 1/16 in	260	25	376/3½	3	.1065	Yes	376/10
B.31	1 in	200	30	376/3½	2	.1065	Yes	376/82
B.31	1 in	260	30	376/3½	2	.1065	No	376/81
B.32	1 in	200	30	376/3½	2	.1065	Yes	376/82
B.32	1 in	260	30	376/3½	2	.1065	No	376/81
B.33	1 1/16 in	210	25	376/3½	3	.1065	Yes	376/84
B.33	1 1/16 in	260	25	376/3½	3	.1065	No	376/85
B.34	1 1/16 in	210	25	376/3½	3	.1065	Yes	376/84
B.34	1 1/16 in	260	25	376/3½	3	.1065	No	376/85

1 General description

1   The fuel system comprises a petrol tank from which petrol is fed by gravity to the float chamber of the carburettor via a petrol tap incorporating a gauze filter. Some competition models are fitted with two petrol taps.

For cold starting the carburettor incorporates an air slide controlled from a lever on the handlebars. This acts as a choke, to temporarily enrich the mixture. As soon as the engine has started, the slide can be raised progressively until the engine will accept full air under normal running conditions. Some models have an air filter attached to the carburettor, to prevent the ingress of dust.

2   A wide variety of carburettors were fitted to the range of BSA 350-500 cc singles, all of which are of Amal manufacture.

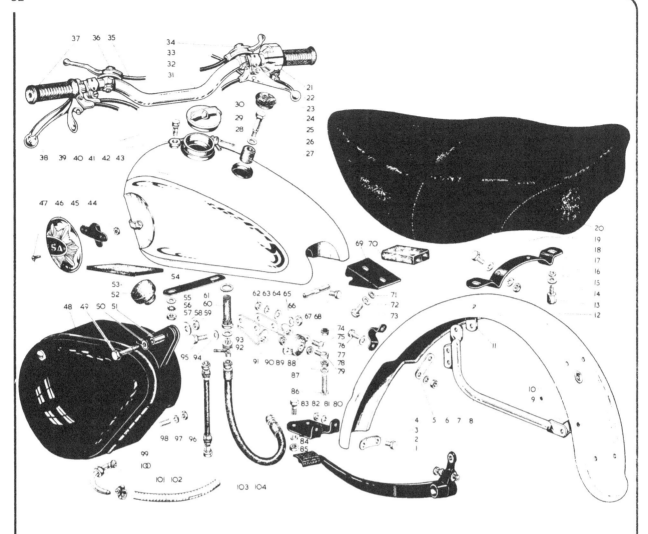

Fig. 4.1. Fuel tank and cycle parts - Gold Star models (other models similar)

1 Rear brake pedal (2 off)	27 Fuel tank	53 Tank mounting rubber	79 Adaptor
2 Bolt	28 Split pin	54 Tank brace plate	80 Mudguard bracket
3 Reinforcing plate	29 Filler cap	55 Washer (2 off)	81 Washer (2 off)
4 Rear mudguard	30 Front brake cable	56 Washer (2 off)	82 Nut (2 off)
5 Reinforcing plate	31 Handlebar	57 Nut (2 off)	83 Bolt (2 off)
6 Spring washer (2 off)	32 Choke cable	58 Washer	84 Washer (2 off)
7 Nut (2 off)	33 Throttle cable	59 Nut	85 Nut (2 off)
8 Mudguard stay (2 off)	34 Choke control	60 Fuel filter	86 Bolt
9 Mudguard bridge	35 Ignition control cable	61 Fibre washer (1 or 2 off)	87 Fibre washer
10 Grommet (2 off)	36 Ignition lever	62 Nut (2 off)	88 Nut (3 off)
11 Mudguard top bridge	37 Left handlebar grip	63 Washer (2 off)	89 Flexible mounting
12 Nut (2 off)	38 Clutch lever	64 Washer (2 off)	90 Nut
13 Bolt (2 off)	39 Valve lifter lever	65 Washer (2 off)	91 Bracket
14 Plain washer (2 off)	40 Valve lifter cable	66 Nut (2 off)	92 Fuel tap
15 Spring washer (2 off)	41 Clutch cable	67 Seat pin	93 Fibre washer (2 off)
16 Spring washer (2 off)	42 Breather screw	68 Bolt	94 Washer
17 Seat bracket	43 Breather pipe	69 Seat bracket (2 off)	95 Bolt (2 off)
18 Washer	44 Nut (4 off)	70 Seat rubber (2 off)	96 Nut (2 off)
19 Bolt (2 off)	45 Clip (2 off)	71 Washer (2 off)	97 Washer (2 off)
20 Dualseat	46 Badge (2 off)	72 Washer (2 off)	98 Stud (2 off)
21 Throttle twistgrip	47 Screw (4 off)	73 Bolt (2 off)	99 Fuel pipe
22 Front brake lever	48 Toolbox	74 Nut	100 Gland nut
23 Centre bolt blanking plug	49 Bolt	75 Bolt (2 off)	101 Hose clip (2 off)
24 Fixing bolt	50 Washer	76 Bolt (2 off)	102 Plastic fuel hose
25 Washer	51 Distance tube	77 Washer (2 off)	103 Petrol pipe
26 Rubber mounting	52 Tank mounting rubber (2 off)	78 Fixing clip (2 off)	104 Petrol pipe

The M series models and the B31-34 models are all fitted with the 'Monobloc' carburettor, the instrument which superseded the separate float chamber standard design. The Gold Star models have also been fitted with the 10TT (TT); Remote Needle (RN) and the Grand Prix (GP) types of carburettor, depending upon the purpose for which the machine was supplied.

Lubrication is effected on the dry sump principle in which oil from the side-mounted oil tank is delivered by gravity to the gear-type pump located in the base of the crankcase. Oil is distributed under pressure to the big-end bearings, via the hollow right-hand end of the crankshaft. Distribution is effected via the timing cover which has internal oilways that match up with the crankcase and a jet for the big-end feed. Separate feeds are directed to the cams and cam followers. In the same manner, a rocker oil feed pipe carries oil from the oil return pipe to the rocker box and valve gear. Surplus oil drains back via the push-rod cover tube into the timing chest and thence into the crankcase via a hole in the crankcase wall (ohv models only).

Other parts of the engine are lubricated by splash from the positive feed.

The lubrication system is protected by two filters and a pressure release ball-valve. The main filter, which is of the gauze or felt type (depending on model), is contained within the oil tank. The secondary filter, also of the gauze type, is fitted above the crankcase sump plate, and prevents larger particles of matter from passing through the return side of the pump.

2 Petrol tank: removal and replacement

1 Turn the petrol tap(s) to the 'OFF' position and disconnect the petrol pipe(s) by unscrewing the union(s) at the base of the tap(s).
2 On M series machines, remove the two bolts which pass through the lugs projecting from the front of the petrol tank, either side of the steering head lug. Unscrew the two bolts that pass into the base of the tank, towards the rear. Before the petrol tank can be lifted away from the machine, the single bolt retaining the front of the dualseat must be removed to enable the seat to be lifted slightly.

On all other models remove the rubber plug from the centre of the tank top and use a box spanner to slacken and remove the retaining bolt. Remove also the metal strip that bridges the two halves of the tank at the front. The bridge piece is held by two

nuts on studs.
3 Replacement is accomplished by reversing the procedure adopted for the removal of the tank. Do not omit to replace the bridging strip fitted to the underside of the tank of the swinging arm models. This strip prevents the two halves of the tank from flexing and opening the welded seams.
4 After replacing the tank, check that the air hole in the filler cap is unobstructed. If the tank is airtight, the supply of petrol will be cut off, leading to a mysterious engine fade out that is difficult to eliminate without realising the cause.

3 Petrol tap: removal and replacement

1 Each petrol tap threads into an insert welded in place at the rear of the tank, and seats on a fibre washer which should be replaced to prevent leakage each time a tap is removed.
2 Several different types of tap have been employed. Most models have a tap of the push-pull type, some of which incorporate a renewable cork plunger. Many competition machines have a tap of the lever type.
3 Taps of the pull-out type are unlikely to give trouble unless the machine is left idle for a long period with the petrol tank dry. Under these conditions the corks will shrink, causing the taps to leak until the corks swell again. Replacement corks were available originally, but it is doubtful whether they can be obtained now. It is sometimes possible to restore a badly shrunken cork by soaking it in castor oil, or if a longer period is allowed, in petrol.
4 Repair of a leaking lever type tap is not practicable unless the lever is detachable. If this is the case the taper on the lever shank may be ground into the tap body, using fine grinding compound.
5 If it is necessary to remove one or both taps, the tank must be drained of petrol first. If the contents of the tank are low, it is sometimes possible to remove and replace a tap without spillage if the machine is laid over at an angle, on one side.

4 Petrol feed pipes: examination

1 The petrol feed pipes are fitted with unions to make a quickly detachable joint at both the carburettor float chamber and the two petrol taps. Leakage is unlikely to occur, unless the tubing cracks or splits.
2 Either synthetic rubber or plastic tubing is used, sometimes

4.1 Petrol feed pipes are fitted with union to aid removal

5.1 All carburettors are flange fitting

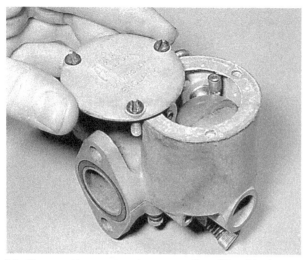

6.1a Remove side cover to gain access to float (Monobloc)

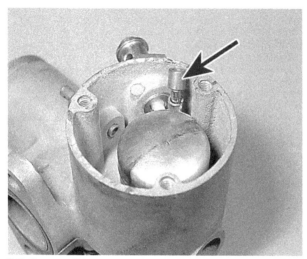

6.1b Note small brass spacer on pivot

6.1c Do not lose tiny nylon float needle

with an outer protective covering of metal braiding. After a long period of service, the plastic tubing will harden, due to the gradual removal of the plasticiser by the petrol. If the pipes are exceptionally rigid, they should be replaced because it is under these conditions that the pipe is most likely to crack. Synthetic rubber pipes rarely give trouble. The chief danger is small particles of rubber breaking away from the inner walls during service, which may pass into the carburettor float chamber and cause the float needle to stick.

3   Never use ordinary rubber tubing, even as a temporary expedient. Petrol causes rubber to swell and disintegrate, thereby completely blocking the fuel supply.

5   Carburettor: removal

1   Irrespective of the type of carburettor fitted a flange-fitting method of mounting to the induction manifold is used.
2   Before removing the carburettor from the two studs that extend from the inlet port, the top of the mixing chamber should be released to enable the slide and needle assembly to be lifted out and taped to a convenient frame member, out of harms way. On the Monobloc carburettors, the circular mixing chamber top is retained by a screwed ring. A similar fitting is employed in the case of the TT, GP and RN racing carburettors, but before the ring can be unscrewed, a locking bolt or a spring clip must be displaced first.
   The choke assembly fitted to GP, TT and RN carburettors is remote from the throttle valve assembly and must be detached from the carburettor by unscrewing the housing nut. The housing and plunger, complete with control cable, can then be withdrawn.
3   It is also necessary to disconnect or remove the air filter (if fitted) depending on which type is used. The smaller type having a perforated case screws directly to the carburettor air intake. The larger type that contains a detachable filter element within a metal box is attached by means of a short length of synthetic rubber hose, which must be pulled off the carburettor air intake. No air filter is fitted to the TT, GP or RN carburettors.

6   Carburettor: dismantling, examination and reassembly

Amal Monobloc carburettor only

1   The Monobloc carburettor is so named, as it was the first instrument manufactured by Amal Limited that incorporated the float chamber as an integral part of the main carburettor body. To gain access to the float and float needle, it is necessary to remove the circular cover that forms the side of the float chamber housing. It is retained by three screws and has a gasket on the inside joint. The float hinges on a pin projecting from the float chamber wall. It is preceded by a small brass spacer. When the float is withdrawn, the nylon float needle will be displaced from its seating and fall clear.
2   The jet block is retained by a screw close to the pilot jet adjusting screw, and by a large hexagon nut at the base of the mixing chamber. Before the latter can be unscrewed, the hexagon bolt below it should be removed first and the main jet and needle jet unscrewed from the jet block. The main jet threads into the lower end of the needle jet, which itself screws into the mixing chamber base. When the large hexagon nut has been removed and the small screw in the side of the mixing chamber, the jet block can be drifted out of position in an upwards direction, using great care because it is made of brass.
3   The throttle slide, needle and air slide assembly will remain attached to the mixing chamber top. If it is necessary to dismantle this assembly, retract the air slides by operating the handlebar control and disengage it from the slot in the top of the throttle slide. Remove the throttle cable from the throttle slide by withdrawing the split pin that retains the lower cable nipple in its seating, then raise the slide upwards against the pressure of the return spring so that the cable can be disengaged completely. Take off the return spring and place it in a safe

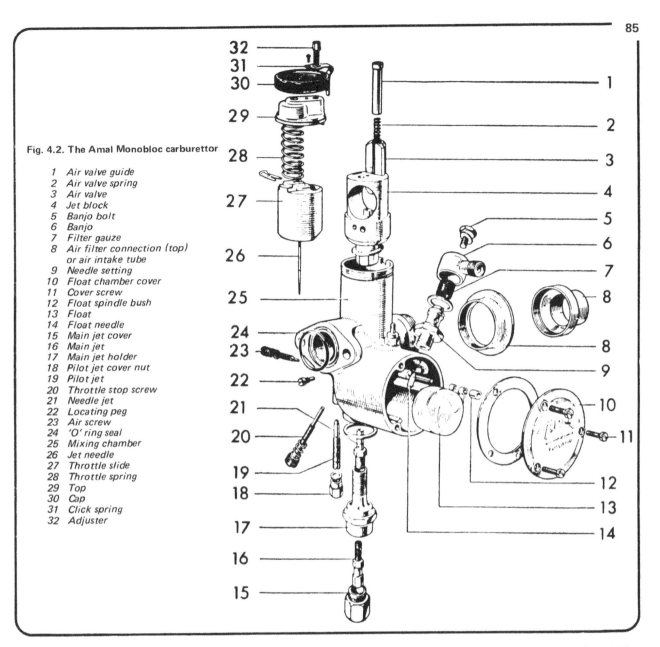

Fig. 4.2. The Amal Monobloc carburettor

1 Air valve guide
2 Air valve spring
3 Air valve
4 Jet block
5 Banjo bolt
6 Banjo
7 Filter gauze
8 Air filter connection (top)
  or air intake tube
9 Needle setting
10 Float chamber cover
11 Cover screw
12 Float spindle bush
13 Float
14 Float needle
15 Main jet cover
16 Main jet
17 Main jet holder
18 Pilot jet cover nut
19 Pilot jet
20 Throttle stop screw
21 Needle jet
22 Locating peg
23 Air screw
24 'O' ring seal
25 Mixing chamber
26 Jet needle
27 Throttle slide
28 Throttle spring
29 Top
30 Cap
31 Click spring
32 Adjuster

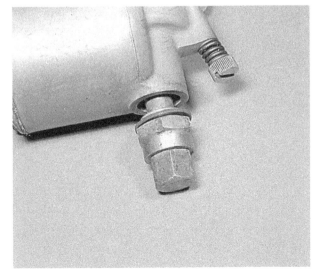

6.2a Remove large bolt which serves as ...

6.2b ... main jet holder in carburettor base

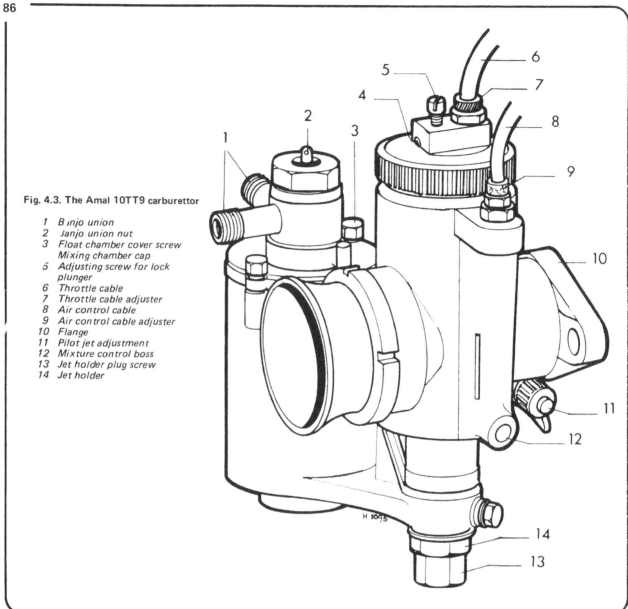

Fig. 4.3. The Amal 10TT9 carburettor

1 Banjo union
2 Banjo union nut
3 Float chamber cover screw
   Mixing chamber cap
5 Adjusting screw for lock
   plunger
6 Throttle cable
7 Throttle cable adjuster
8 Air control cable
9 Air control cable adjuster
10 Flange
11 Pilot jet adjustment
12 Mixture control boss
13 Jet holder plug screw
14 Jet holder

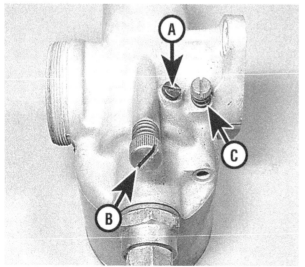

6.2c A, jet block screw; B, throttle stop screw; C, pilot air screw

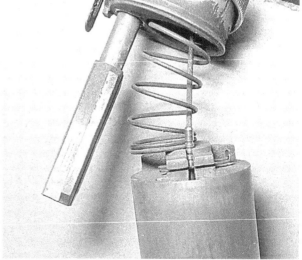

6.3a Jet needle retained by clip. Remove clip to ...

place for reassembly.

4 The needle is retained by a spring clip. Before withdrawing the clip to release the needle, note the needle position. It has five notches to give variation of mixture strength and must be replaced in the same position.

5 It is unlikely that the air slide assembly will need to be dismantled. If, however, such action is necessary, displace the lower cable nipple from its seating in the base of the slide, so that it protrudes through the slot in the slide body. The slide can be pulled off the cable, followed by the shouldered guide on which it seats, when raised. Do not misplace the return spring.

6 If it is necessary to dismantle the float chamber tickler, access is gained by unscrewing the hexagon nut that surrounds the tickler plunger. The Monobloc carburettor has a detachable pilot jet. It is housed in the underside of the mixing chamber body, close to the flange joint and is blanked off by a hexagon headed plug. If this plug is removed, the threaded pilot jet within the carburettor body can be unscrewed. It has a slotted end, to facilitate removal with a screwdriver.

Amal 10TT carburettor only

7 This type of carburettor has many similarities with the Monobloc carburettor, especially with regard to the location of the jet block and the arrangement of the jets. A separate float chamber is fitted and the air slide is carried in a separate compartment cast on the right-hand side of the mixing chamber. The most significant difference is that the pilot jet screw meters the flow of petrol and not air as in the previously mentioned carburettor. This means the adjustment works in the opposite direction; when the adjusting screw is moved inwards, it weakens and not enriches the mixture. The pilot jet is detachable but not variable in size.

Amal GP2 carburettor only

8 The GP2 carburettor is similar in many respects to the 10TT type, although in this instance the casting that contains the air slide assembly is on the left-hand side of the mixing chamber. The most noticeable difference is the use of a remote mounted float chamber, which must be accurately positioned so that the fuel level is correct.

9 The float chamber, and hence the fuel level, is correct when the scribe mark on the float chamber is exactly level with the bottom of the circle scribed on the head of the blanking plug that is diametrically opposite the pilot adjustment screw. Early carburettors do not have a scribe mark on the float chamber, which should be exactly 1 inch below the float chamber cap mating surface.

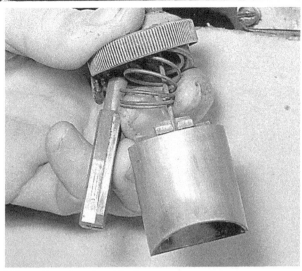

6.3b ... allow throttle cable disconnection

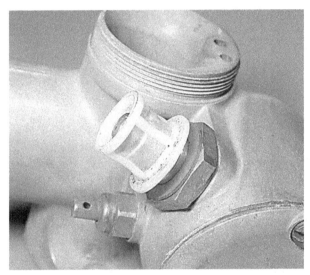

6.6a Filter incorporated in petrol feed

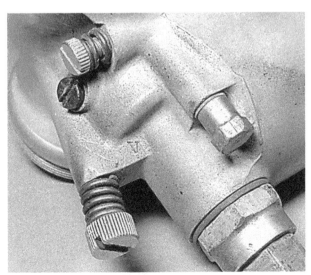

6.6b Remove sleeve nut and fibre washer to ...

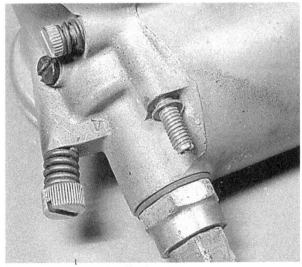

6.6c ... gain access to the pilot jet.

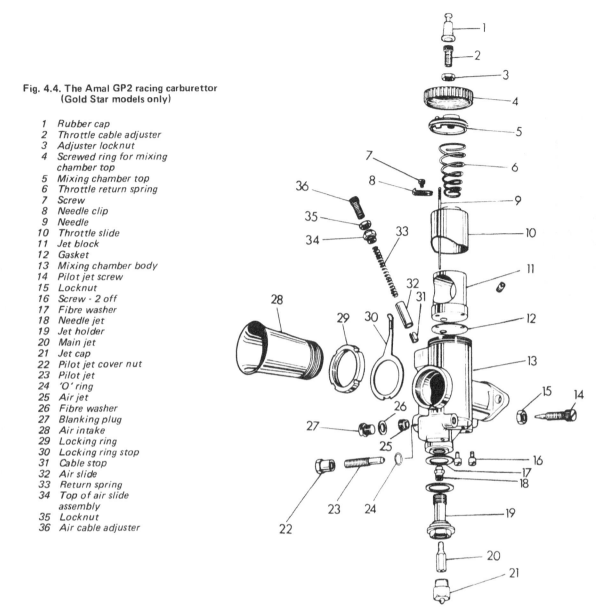

**Fig. 4.4. The Amal GP2 racing carburettor
(Gold Star models only)**

1   Rubber cap
2   Throttle cable adjuster
3   Adjuster locknut
4   Screwed ring for mixing
    chamber top
5   Mixing chamber top
6   Throttle return spring
7   Screw
8   Needle clip
9   Needle
10 Throttle slide
11 Jet block
12 Gasket
13 Mixing chamber body
14 Pilot jet screw
15 Locknut
16 Screw - 2 off
17 Fibre washer
18 Needle jet
19 Jet holder
20 Main jet
21 Jet cap
22 Pilot jet cover nut
23 Pilot jet
24 'O' ring
25 Air jet
26 Fibre washer
27 Blanking plug
28 Air intake
29 Locking ring
30 Locking ring stop
31 Cable stop
32 Air slide
33 Return spring
34 Top of air slide
    assembly
35 Locknut
36 Air cable adjuster

Inspection - all models

10 Check the float to see whether it has become porous and allows petrol to enter and upset its balance. Irrespective of whether the float is made of copper or plastics, it should be replaced if a leak is evident. It is not practicable to effect a satisfactory repair.

11 Check the float needle and float needle seating to see whether the float needle is bent or whether a ridge has worn around either the needle or its seating as the result of general wear. All defective parts should be replaced. The needle seat will unscrew from the float chamber body, to permit replacement when necessary.

12 The throttle slide, needle and air slide assembly still attached to the carburettor top should be examined. Signs of wear on the throttle slide will be self-evident, if the amount of wear is particularly high it may be responsible for a pronounced clicking noise when the engine is running slowly, as the slide moves backwards and forwards within the mixing chamber. Check also the small projection within the lower edge of the throttle slide on machines fitted with the 10TT and other racing carburettors. If this projection wears away or is broken off, there is nothing to prevent the slide from turning and causing the throttle to stick open.

13 The needle should be straight and the needle retaining clip a good fit. Check the needle for straightness by rolling it on a sheet of plate glass. If it is bent, it must be replaced. Reject any needle clip that has lost its tension.

14 The air slide assembly seldom requires attention. Trouble can occur if the compression spring loses its tension, since this will cause the air slide to stick, making cold starting more difficult.

15 Check the main jet, needle jet and pilot jet (if fitted). Wear will occur in the needle jet only; the other jets are liable to blockages if dirty or contaminated petrol is used. NEVER use wire to clear a blocked jet otherwise there is danger of the hole being enlarged. Use either a foot pump or a compressed air line to clear the blockage.

16 The 'Monobloc' carburettor fitted to some models has a synthetic rubber 'O' ring in the centre of the mounting flange. This seal must be in good condition to prevent air leaks.

17 On all carburettors there is a pronounced tendency for the mounting flange to 'bow' if the retaining nuts are overtightened. The resultant air leak, which will have a marked effect on carburation, will be difficult to trace as a result. The condition of the flange can be checked by holding a straight edge across the face. If the flange is bowed, it should be rubbed down with

a sheet of fine emery cloth wrapped around a piece of flat glass, using a rotary motion, until the bow is removed. Make sure the carburettor body is washed very thoroughly after this operation, to ensure that small particles of abrasive do not lodge in any of the small internal air passages.

All carburettors

18   To reassemble the carburettor, reverse the procedure used for dismantling. Note that undue force should never be used during dismantling or reassembly because the castings are usually in a zinc-based alloy which will fracture very easily if overstressed.
19   The 10TT, RN and GP2 racing carburettors have provision for locking wires through certain bolts and it is advisable to replace the locking wire when reassembly is complete. There is a greater tendency for parts to work loose and become lost at high speeds, as the result of vibration. Locking wire will ensure this cannot happen.

7   Carburettor: checking the settings

1   The various sizes of jets and that of the throttle valve and needle are predetermined by the manufacturer and should not require modification. Refer to the Specifications list if there is any doubt about the valves fitted. It should be noted that carburettor settings on all competition machinery should be made in reference to particular combinations of exhaust pipe length and silencer type.
2   Slow running is controlled by a combination of the throttle stop and air regulating (pilot jet) screw settings. Commence by screwing the throttle stop screw inwards so that the engine runs at a fast tickover speed. Models fitted with the Amal 10TT, RN and GP2 carburettors do not have a throttle stop screw and in consequence this adjustment must be made with the throttle cable adjusting screw. Adjust the air screw setting until the tickover is even, without either misfiring or hunting. Readjust the throttle stop (or cable adjuster screw) until the desired tickover speed is obtained, then re-check with the air regulating screw so that the tickover is as even as possible. Remember that turning the pilot adjuster screw out on Monobloc carburettors weakens the mixture, but enriches the mixture on the GP, TT and RN instruments. Always make these adjustments with the engine at normal running temperature and remember that the more sporting models fitted with a racing carburettor and having a high degree of valve overlap are unlikely to run very evenly at low speeds, no matter how carefully the adjustments are made. In this latter category, it is often advisable to arrange the throttle to shut off completely when it is closed, so that maximum braking effect can be obtained from the engine on the over-run.
3   As an approximate guide, up to 1/8 throttle is controlled by the pilot jet, from 1/8 to ¼ throttle by the throttle slide cutaway, from ¼ to ¾ throttle by the needle position and from ¾ to full throttle by the size of the main jet. These are only approximate divisions: there is a certain amount of overlap.
4   Note that machines fitted with an air filter have a slightly smaller size of main jet than standard. This is to compensate for the reduced air flow, which tends to make the mixture richer. It follows that when an air filter is disconnected, the larger size main jet must be fitted. If it is not a weak mixture will result with the possibility of engine damage through overheating.
5   It cannot be overstressed that the correct float level height is essential if machines fitted with the GP2 carburettor are to perform satisfactorily. If the float chamber is either too high or too low, carburation throughout the entire throttle opening range will be affected.

8   Air filter: location, examination and cleaning element

1   The air filter fitted as original equipment to some models is contained within a box attached to the battery mounting

strap and is of C and W manufacture.
2   The element can be removed after detaching the air filter box and removing the gauze screen which is retained by screws. The element should be cleaned by washing with petrol, then it should be immersed in thin oil (SAE 20) and allowed to drain thoroughly before it is re-inserted in the filter housing.
3   A non-standard air filter, often fitted after manufacture, has an element which is contained within a perforated metal box that screws directly on to the carburettor air intake. The internal element is retained by a single screw. It is cleaned by adopting the technique used for the filter element in the preceding paragraph.
4   Under normal conditions, the air filter requires attention about every 2,000 miles. If the machine is used in a particularly dusty situation, this period should be halved.
5   If the air filter is detached for any reason, it is imperative that the carburettor is re-jetted to compensate for the changes in carburation that will occur. This means INCREASING the size of the main jet by at least 10; a machine fitted with an air filter has a smaller than standard main jet fitted, to compensate for the restriction of air flow imposed by the air filter and the consequent enriching of the mixture.
6   The air filter hose (if fitted) must be in good condition if the filter is to function effectively. Replace any hose that shows signs of either cracking or splitting.

9   Exhaust systems

1   The exhaust system rarely requires attention because it does not need a regular clean-out like that of a two-stroke. It is important that both the exhaust pipe and silencer are rigidly mounted so that they cannot work loose, and that there is no air leak at any point in the system. An air leak will give rise to an elusive backfire when the engine is on the over-run and can prove difficult to trace.
2   If possible, when renewing the silencer, fit a genuine BSA part to retain the original design characteristics. Failing that, a good pattern part must be fitted. Especially on non-competition models avoid fitting unsuitable silencers as the change in design will inevitably alter the breathing of the engine, requiring an alteration in main jet sizes and also the throttle slide. Arriving at the correct carburettor settings to meet the requirements of an altered exhaust system requires skill and experimentation that is generally outside the scope of the average owner.

10   Engine lubrication

1   A dry sump lubrication system is employed, in which oil is drawn from an oil tank mounted on the right-hand side of the machine by a double gear pump located in the crankcase. The oil pump is driven from the right-hand end of the crankshaft assembly and has two sets of gears, one pair of which form the feed pump and the other the scavenge pump. It is the function of the scavenge pump to return oil to the oil tank after it has circulated through the engine. For this reason, the gears of the scavenge pump are wider so that the scavenged return feed can never be overtaken by the rate of delivery from the feed pump.
2   The feed rate is constant and no means is provided for either reducing or increasing the rate of delivery. Because oil can pass through the teeth of a gear-type pump whilst the engine is stationary, it is necessary to incorporate some device in the feed line to cut off the flow of oil immediately the engine stops. This takes the form of a non-return ball valve which lifts from its seating immediately the engine starts.
3   The lubrication system functions as follows. Oil from the feed pump outlet passes through a drilling in the crankcase to the timing cover, which acts as the distributor for the oil supply. One feed is direct to the big-end bearing, through a crankshaft oil jet, sometimes referred to as the quill, and through the hollow right-hand crank. A second feed is taken from the oilways in the timing chest and feeds the cams and cam followers.

The third feed is taken from the oil return pipe and is delivered to the rocker box by means of an external delivery pipe (ohv models only).

4 Engine components not directly lubricated receive their oil quota as the lubricant drains back into the crankcase. The piston, cylinder and small end receive oil dissipated by the rapidly revolving big-end assembly, which also keeps the main bearings constantly supplied. A drilling in the crankcase wall diverts oil to the idler pinion bush and there is a drain back into the crankcase from the base of the timing chest. Oil that collects within the crankcase is picked up by the scavenge pump and returned to the oil tank. Impurities or foreign bodies are prevented from passing through the return side of the pump by a gauze screen fitted above the sump plate. By removing the filler cap from the top of the oil tank, the oil return can be checked visually.

11 Lubrication: maintenance

1 Always check the oil return to the oil tank before setting out. If the filler cap is removed, the oil return can be seen from the return pipe immediately below the orifice. If no return is evident immediately the engine starts (except after a rebuild), stop the engine and investigate the cause. Most probably the anti-syphon valve in the crankcase is sticking. To check, remove the sump plate that is retained by four nuts and spring washers and insert a piece of wire into the orifice of the pipe so that the ball bearing within is raised from its seating. Replace the cover and filter, making sure the pipe passes through the hole in the filter gauze, before the engine is restarted. If there is still no return, unscrew the pressure release valve a few threads and re-start the engine. If oil is evident around the threads, this will prove the main feed section of the oil pump is working.

2 The evidence of air bubbles in the returned oil does not indicate a fault. The scavenge pump has a greater capacity than the feed pump and after the initial excess of oil has been pumped away from the crankcase, into which it has drained, the pump will work at reduced capacity and air will be included in the pick-up.

3 Do not omit to change the engine oil at the recommended intervals. Whilst the tank is drained, the opportunity should also be taken to clean all the filters in the system by removing them and washing them in petrol before they are replaced.

4 Leakage from the filler cap of the oil tank often indicates that the pressure relief pipe is blocked. Check also the oil return pipe. On early models the pipe extends from the base of the oil tank. Later machines have a froth tower on top of the tank to

which the pipe is attached. In either case the pipe can be unblocked, using a length of stiff wire.

5 The timing chest/crankcase breather is unlikely to give trouble unless the pipe becomes trapped or blocked. On some Gold Star models the breather is timed, the valve piece being driven from the magneto pinion. It is imperative that the magneto pinion is positioned correctly during ignition timing, if the breather is to function effectively.

12 Oil filters: cleaning

1 All models are fitted with two filters. One of which is situated in the oil tank and the other in the crankcase.

2 On M series machines, the filter can be removed from the top of the oil tank by removing the filter seating from below the oil filler cap. On all other models, the filter is an integral part of the drain plug and consists of a mesh cylinder. Both filters can be washed in petrol. The M series filter has a heavy wire matrix and will require immersing in petrol to ensure good cleaning. Both filters should be allowed to dry before re-use.

3 The crankcase oil filter consists of a perforated brass sheet which prevents large particles of matter from blocking the return pipe non-return ball valve, or passing through the scavenge side of the oil pump. Access to and removal of the filter can be made after removing the sump plate which is retained on four studs by nuts and washers. Before removing the sump plate, remove the drain plug from the base of the crankcase and allow the oil trapped in the crankcase to drain off.

13 Removing, dismantling and reassembling the oil pump

1 The oil pump should not be disturbed or dismantled unless absolutely necessary. It will give long service without trouble unless mechanical failure within the engine causes particles of debris to pass through the teeth of the pinions.

Progressive wear of the pump teeth and an increase of gear endfloat may eventually give rise to a slow rate of oil delivery and return. A sluggish rate of return is often accompanied by heavy oil loss from the crankcase breather, and a spate of oil leaks due to the increased amount of oil permanently trapped in the crankcase.

2 Access to the pump is gained after draining the oil within the crankcase and after removal of the sump plate and filter screen. The pump is retained in position by two of the four bolts which pass through the pump. The two bolts may be

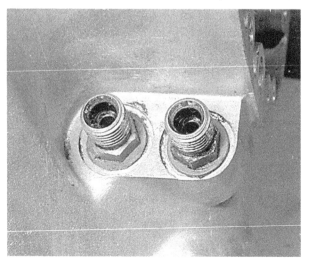

13.1a Oil feed and return unions are sealed by fibre washers

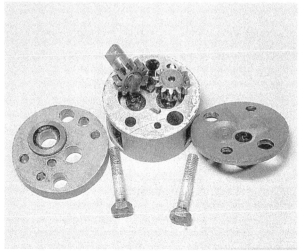

13.1b Oil pump - component parts

identified by the spring washer below each bolt head. After removal of the bolts, pull the pump from position.

3   To dismantle the pump, remove the two remaining bolts and carefully lift off the lower plate, to expose the two scavenge pump gears. Lift off the pump base plate to expose the main pump gears, one of which incorporates a drive tongue which engages with the oil pump spindle. Note the washer which fits on the driven gear boss.

4   Before proceeding further, mark all the loose pinions so that they are eventually replaced in their original locations. It is possible to replace the pinions in an inverted position, which is not advisable since they will have bedded down with their counterparts on the drive spindle.

5   For the oil pump to function efficiently, there must be minimum clearance between the tops of the teeth of each pinion and the pump body. They must also revolve quite freely when the pump is fully assembled. Pinions with chipped or fractured teeth must be renewed.

6   Refitting of the pump may be carried out by reversing the dismantling procedure. Always use a new pump base gasket.

13.6a Always use new oil pump gasket

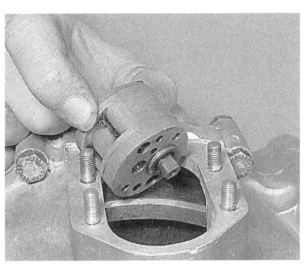

13.6b Pump drive dog must engage with slot in driveshaft

14 Fault diagnosis: fuel system, carburation and engine lubrication

Symptom	Cause	Remedy
Engine 'fades' and eventually stops	Blocked air hole in filler cap	Clean.
Engine difficult to start. Fuel drips from carburettor	Carburettor flooding	Dismantle and clean carburettor. Check for punctured float.
Engine runs badly. Black smoke from exhaust	Carburettor flooding	Dismantle and clean carburettor. Check for punctured float.
Engine difficult to start. Fires only occasionally and spits back through carburettor	Weak mixture	Check for fuel in float chamber and whether air slide working.
Oil does not return to oil tank	Sticking anti-syphon valve	Stop engine immediately. Check anti-syphon valve in sump plate.
Engine joints leak oil badly	Pressure release valve sticking Crankcase breather timed incorrectly	Dismantle and clean valve assembly. Check marks on timing pinions for correct alignment.
Oil consumption heavy. Blue smoke from exhaust	Engine in need of rebore	Rebore cylinder and fit O/S piston.

Chapter 5 Ignition system

Contents

General description 1
Removing and replacing the magdyno 2
Magneto: examination and maintenance 3
Contact breaker: adjustment 4
Contact breaker points: removal, renovation and replacement 5
Condenser: faults and location 6
Replacing the magneto pinion oil seal and a drive gear mesh 7
Magneto overhauls and repairs 8

Ignition timing: checking and resetting 9
Contact breaker adjustment: alternator models only ... 10
Contact breaker: removing, cleaning and replacement ... 11
Auto-advance unit (alternator models only) 12
Ignition timing: checking and resetting (alternator models only) 13
Condenser: faults (alternator models) 14
Ignition coil: checking 15
Spark plugs: checking and resetting the gap 16
Fault diagnosis 17

Specifications

M series and B series, except alternator models

Magneto

Make	Lucas
Type	M01/6 (045160 or 045232)
Drive	Anticlockwise
Contact breaker gap	0.012 in (0.30 mm)
Spark plug:	
B31, B33, M33	Champion L - 7 or L - 85, or Motorcraft AE2
B32	Champion N - 4 or Motorcraft AG2
B34	Champion N - 5 or Motorcraft AG3
M20, M21	Champion N - 8 or Motorcraft AG5

Gold Star models (touring, clubmans and racing)

Magneto

Make	Lucas
Type	Magdyno FN01/6 (046504)
Drive	Anticlockwise
Contact breaker gap	0.012 in (0.30 mm)
Spark plug	Champion N57R or Motorcraft AG603 (racing)
	Champion N3 or Motorcraft AG-1 (touring)
	Champion N54R or Motorcraft AG403 (clubmans)

Gold Star models (Scrambles)

Magneto

* Make	Lucas or BTH
Type	KMR1 (42123L) or KD1
Drive	Anticlockwise
Contact breaker gap	0.012 in (0.30 mm)
Spark plug	Champion N3 or Motorcraft AG1

* Either make of magneto was also fitted to Clubmans and Racing models as an option

B31 and B33 - 1958 on

Alternator

Make	Lucas
Type	Six coil, rotating permanent magnet
Output	6 volts - 60 watts

Ignition timing

M20, M21, B31, B32, B34 and pre 1956 B33	7/16 in (11.0 mm) BTDC fully advanced
M33 and Post 1955 B33	3/8 in (9.5 mm) BTDC fully advanced

*Gold Star models *

CB34GS, DB32GS,

Touring	15/32 in (11.8 mm) (39°) BTDC fully advanced
Scrambles	15/32 in (11.8 mm) (39°) BTDC fully advanced
Racing	15/32 in (11.8 mm) (39°) BTDC fully advanced.

DB34GS and DBD34GS

Clubman's	...	...	...	...	...	...	...	...	15/32 in (11.8 mm) (39°) BTDC fully advanced

CB34GS and DB34GS

Touring (CB34GS only)		...	...	...	...	.....	...	13/32 in (10.3 mm) (36°) BTDC fully advanced
Scrambles	...	...	...	...	...	...	...	...	13/32 in (10.3 mm) (36°) BTDC fully advanced
Racing	...	...	...	...	...	..	...	...	13/32 in (10.3 mm) (36°) BTDC fully advanced
Touring (DB34GS only)		...	...	...	...	...	...	½ in (12.7 mm) (41°) BTDC fully advanced

All ignition timing given is for machines having standard compression ratios and running on petrol (touring) or petrol and Petrol/Benzole fuel (competition machinery)

---

1 General description

On all models except the 1958-60 B31-33 machines, a magneto is used to generate the spark necessary for igniting the mixture at the correct time, when the cylinder is on the compression stroke. The magneto is a self-generating instrument that does not depend on a battery for the initial electrical voltage. This voltage is developed in the primary windings of the armature as it revolves in close proximity to the pole pieces of a magnet. By interrupting the electrical circuit at the appropriate time, using a contact breaker, a very high voltage is developed in the secondary windings of the armature that causes a spark to jump the air gap between the points of the spark plug.

The magneto is driven from the timing gears and is timed so that each spark will occur at the precise time when it is most necessary to fire the mixture that is under compression in the cylinder. Provision is made to either advance or retard this setting as the engine speed increases or decreases, so that the spark across the points of the spark plug can be used to maximum advantage.

Because the magneto is a self-generating instrument, the machine can still be used when the battery is detached, or if the electrical system is not in working order, although statutory requirements would make such use illegal. The chief disadvantage of the magneto is that it develops a low voltage at low rotary speeds and in consequence the spark is less intense, which sometimes makes starting more difficult. In practice, such problems rarely occur provided the magneto is maintained in good working order.

The magneto is now classified as an obsolete instrument, and all production in the UK has ceased. The corresponding shortage of spares makes it necessary to take a defective magneto to a magneto repair specialist, who will either renovate the existing instrument or replace it with a service exchange magneto that has been reclaimed at an earlier date.

On all models, except some Gold Star machines, the magneto is incorporated in a Magdyno instrument where a dynamo is strapped to the main body of the instrument. Some Gold Stars intended only for competition dispense with the dynamo and utilise a racing magneto only.

From 1958, the B31 and B33 models were fitted with an alternator in place of the dynamo. In addition to this, the magneto was dispensed with and coil ignition fitted, using a contact breaker assembly driven direct from the timing gears.

2 Removing and replacing the Magdyno

1 Removal of the Magdyno is not necessary unless attention to the dynamo drive clutch is required or the magneto armature is to be removed. Routine attention to the contact breaker assembly or to the HT pick-up can be made with the Magdyno in position.

2 Before the Magdyno is removed, the magneto pinion must be withdrawn from the shaft after removal of the timing chest cover. Because the pinion is a tapered fit on the magneto armature shaft, a special puller is required to draw the pinion from the shaft. The correct tool is BSA part No. 61-1903. Loosen and remove the pinion retaining nut and washer. Screw the special puller into the threaded portion of the pinion and then tighten down the centre bolt until the pinion is freed.

3 Disconnect the advance-retard cable from the handlebar control. In order to provide enough slack in the cable to facilitate disconnection, it may be necessary to detach the cable at the magneto, removal will prevent it becoming snagged. Unscrew the cable housing and pull the cable so that the nipple is exposed. The nipple can then be slid from position in the cam rotating plunger.

2.2 Magdyno is retained by a strap and bolt

2.3 Disconnect the manual ignition control cable

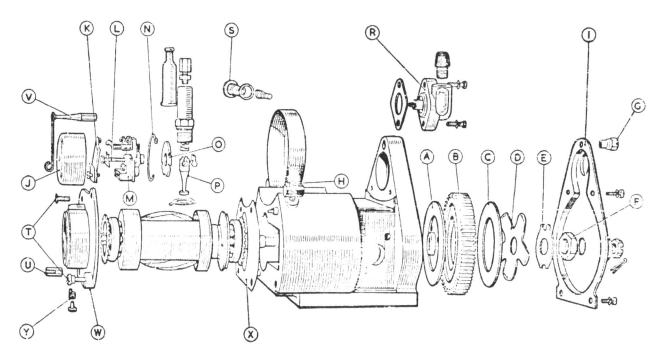

Fig. 5.1. Exploded view of typical single-cylinder Magdyno

a	Gear mounting plate	h	Dynamo strap	m	Contact breaker body	t	End plate screws
b	Fibre driving gear	i	End cover	n	Wire clip	u	Earth terminal
c	Pressure ring	j	Contact breaker cover	o	Face cam	v	Pillar
d	Pressure spring	k	Contact breaker moving	p	Plunger	w	End plate
e	Locking plate		point	r	Magneto pickup	x	Shim
f	Nut	l	Contact breaker screw	s	Earth brush assembly	y	Stop
g	Nut						

4   Unscrew and remove the strap bolt retaining the magneto in position. On some models removal of the carburettor is necessary, to give clearance when unscrewing the bolt. Lift the Magdyno from position.

5   Note any shims placed between the base of the Magdyno and the engine platform which are fitted to adjust the magneto pinion/idler pinion mesh.

6   The Magdyno may be refitted by reversing the dismantling procedure. If mesh adjustment shims were fitted, they must be replaced to ensure that the correct mesh is maintained. Before tightening the retaining strap, push the instrument firmly up against the rear of the timing chest.

7   After refitting the Magdyno, the ignition timing should be carried out as described in Section 9 of this Chapter.

3   Magneto: examination and maintenance

1   With the exception of the contact breaker assembly, as discussed in the following section, the only parts of the magneto likely to require attention are the pick-up brush and the slip ring, which should be checked at regular intervals. There is no necessity to remove the magneto from the machine for this type of maintenance.

2   Remove the pick-up brush holder from the left-hand end of the magneto by removing the retaining screws. The brush holder will lift out, complete with the pick-up brush. Access is now available to the slip ring. Note the pick-up holder gasket.

3   The slip ring becomes contaminated with carbon dust from the brushes and oil and grease from the bearings. To clean it, wrap a small piece of rag soaked in petrol around a stick, so that the rag can be pushed on to the brass insert of the slip ring as the magneto is rotated. A pencil is often used for this purpose, with the blunt end downwards, but avoid contact with the point

3.2 Pick-up brush holder is retained by two screws

because the lead will act as a pick-up and convey the high tension voltage to the holder! It may be necessary to use a mirror to check when the slip ring is clean once again.

4   Visual inspection will show whether the slip ring is cracked or broken. If either fault is detected, the armature complete with slip ring must be replaced. Water will lodge in a crack or broken portion and cause the high tension voltage to track, leading to a misfire that is very difficult to trace.

5   The brush holder should be cleaned and the brush checked to ensure it is not too short so that good contact with the slip ring is no longer maintained. Check that the brush holder has a cork sealing gasket to exclude water and oil at the flange joint, and a rubber grommet around the spark plug lead to form an effective seal at this point.

6   When replacing the brush holder make sure that the brush is inserted in the tunnel and does not bind. It is possible for the brush to slide out of the tunnel if the holder is replaced carelessly, thereby preventing the brush from contacting the slip ring.

7   If a rubber spark plug lead is fitted, check it for cracks or other surface defects that can cause the spark to jump to a nearby frame or engine component. This form of electrical 'leakage' is most likely to occur when the atmosphere is damp. Plastic-coated cables are recommended, to obviate this trouble.

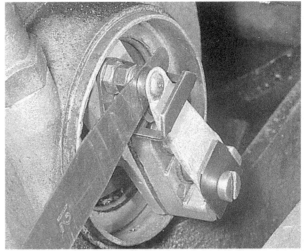
4.1 Points must be FULLY OPEN when checking gap

4 Contact breaker: adjustment

Magneto models only

1   The contact breaker assembly is located behind the detachable end cover at the extreme left-hand end of the magneto. Adjustment is correct if the gap between the contact breaker points is 0.012 in (0.30 mm) when they are fully open.

2   Before checking or adjusting the points gap, examine the surfaces of both points whilst they are in the fully open position. If they are pitted and burnt it will be necessary to remove them for further attention, as described in the following section of this Chapter.

3   Adjustment is effected by unscrewing the locknut of the fixed contact point and either raising or lowering the point until a 0.012 in (0.30 mm) feeler gauge is a good sliding fit between the fixed and the movable points. It is essential that the points are in the fully open position whilst the gap is checked or reset.

4   Before replacing the end cover, give the cam ring a thin smear of grease, taking care none reaches the surface of the contact breaker points. Check that the air hole in the contact breaker end cover is unobstructed, so that the assembly can 'breathe'.

5.2a Remove moving point fixing screw to ...

5 Contact breaker points: removal, renovation and replacement

Magneto models only

1   If the contact breaker points are burnt, pitted or badly worn, they should be removed for dressing. If, however, it is necessary to remove a substantial amount of material before the faces can be restored, they should be replaced.

2   To remove the contact breaker points, first detach the complete contact breaker assembly, which is held on the extreme end of the armature by a centre bolt. To gain access to the bolt head it will be necessary first to remove the contact breaker moving points arm and backing spring which is retained by a single screw. Withdraw the bolt and pull the contact breaker assembly off the shaft. Note that the underside of the contact breaker plate has a horse-shoe shaped projection which registers with a corresponding cutaway in the armature shaft. This will ensure the contact breaker is replaced in exactly the same position, thereby eliminating the need to retime the engine.

3   The fixed contact can be unscrewed from the contact breaker body after loosening the locknut. Removal of the fixed point will free the spring arm guard.

5.2b ... remove assembly retaining screw; note brass tab washer

5.2c Contact breaker assembly - complete

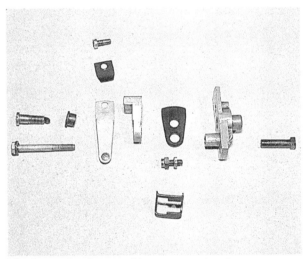

5.2d Contact breaker assembly - component parts

4   When removing the points note the fibre rod which passes through the contact breaker body and transmits movement from the cam to the moving point spring arm. This should be renewed if it has worn to such an extent that movement of the points is limited.

5   The points should be dressed with an oilstone or fine emery cloth. Keep them absolutely square during the dressing operation, or they will make angular contact when reassembled and will quickly burn away.

6   Replace the contact breaker points by reversing the dismantling procedure.

7   Before replacing the points check that the manual advance mechanism is free. Apply a few drops of oil to the plunger and periphery of the cam, if necessary. Lubrication of the cam face is accomplished by means of a wick. Access to the wick is made by removing the screw hidden below the moving point.

8   The cam may be removed for cleaning by prising out the wire retaining clip and allowing the cam to fall from position.

9   Before the machine is used, check that both contact breaker points have an oil and grease free surface. Quite small electrical currents pass between the points and oil or grease will form an effective insulator.

10 It is advisable to clean the contact breaker assembly every 5,000 miles and to check the points gap every 2,000 miles.

11 Some Gold Star models were fitted with a racing magneto in place of the Magdyno unit. Although these magnetos utilise a ring cam operated contact breaker assembly the principles of operation are similar.

6   Condenser: faults and location

Magneto models only

1   A condenser is included in the circuitry of the contact breaker assembly to prevent the points from arcing when they separate. If for any reason the condenser fails, arcing across the points will commence, giving them the characteristic burnt appearance. The most noticeable effect will be a persistent engine misfire, because the spark at the sparking plug will be less intense. The machine will also be much more difficult to start.

2   The condenser is located within the armature. If the condenser is at fault, it is beyond the means of the average rider to effect a satisfactory repair because the armature will have to be stripped. This type of repair should be entrusted to a magneto repair specialist.

5.5 These point faces require dressing

5.8 Face cam is held by retaining clip

7 Replacing the magneto pinion oil seal and adjusting drive gear mesh

1 An oil seal is fitted into a recess in the timing chest, through which passes the boss on the magneto pinion. If the seal is worn, oil will escape from the rear of the timing chest.

2 The old seal can be drifted from position and a new component fitted. Take care not to damage the new seal as it is driven into place.

3 An elusive whining noise coming from the engine can often be traced to the Magneto pinion and idler pinion being in too close mesh with one another. This often happens if a replacement Magdyno is fitted.

4 To check the mesh, drive out the magneto shaft oil seal from the rear of the timing chest and temporarily refit the Magdyno and magneto pinion. The Magdyno securing strap bolt must be tightened fully. Check the backlash between the two gears. There should be just a little play between the two gears at any given position in the rotation of the gears. It should be possible to slide the magneto pinion from position.

5 The correct amount of play is arrived at by placing one or more shims between the Magdyno base and the support platform. If the correct shims are not available, suitable alternatives can be fabricated from brass sheet shim stock. Shim stock of 0.005 in (0.13 mm) thickness is ideal.

6 After adjustment of backlash, remove the Magdyno and refit the oil seal.

8 Magneto overhauls and repairs

1 Although a magneto will normally give a long, reliable service, occasions will occur when repairs are necessary such as the replacement of the armature bearings, the driveshaft oil seal or even an armature rewind. Because the magneto is now classified as an obsolete instrument, spare parts are becoming increasingly difficult to obtain. In consequence, the average rider has little option other than to entrust any such repair work to a magneto repair specialist.

2 Even if the parts required can be located, it is questionable whether the average rider will have sufficient experience to effect a satisfactory repair. For these and other reasons, it is suggested that all such repair work should be carried out by a recognised magneto repair specialist, or alternatively, the faulty magneto replaced with another of identical type. Note that the direction of drive is always described as viewed from the drive end. Some repair specialists may offer a service exchange scheme.

9 Ignition timing: checking and resetting

Magneto models only

1 If the ignition timing is correct, the contact breaker points will be about to separate when the piston is the following distance from top dead centre, on the compression stroke, with the ignition fully advanced:

M20 and M21	7/16 in (11.0 mm) BTDC
B31, B32 and B34	7/16 in (11.0 mm) BTDC
B33 and M33	3/8 in (9.5 mm) BTDC

Gold Star models

CB32 and DB32	15/32 in (11.8 mm) BTDC 39°
CB34 and DB34 (except	
DB34 Touring)	13/32 in (10.3 mm) BTDC 36°
DB34 Touring	½ in (12.7 mm) BTDC 41°
DB34 and DBD34	
Clubmans	15/32 in (11.8 mm) BTDC 39°

2 On all models the ignition timing is given with the piston in the fully advanced position. Timing should always be carried

out with the manual advance-retard cable in the slack position, assuming this is the full advance position.

3 To find top dead centre (TDC) and then the correct piston position, remove the spark plug and insert a short length of welding rod or knitting needle so that it touches the piston crown. On side-valve models a special plug is provided in the cylinder head, directly above the piston, for this purpose. TDC must of course be found on the compression stroke.

4 Mark the rod at TDC so that it corresponds with a suitable fixed mark on the cylinder head. A straight edge placed across the fins is ideal. Make another mark on the rod which corresponds with the piston advance given, and turn the engine backwards past the marked point and then forwards very slowly until the mark is in the correct position with relation to the fixed mark.

5 It is at this point that the contact breaker should be on the verge of opening. This can be verified by retarding the ignition and placing a cigarette paper (or similar thin strip of paper) between the points. Slowly advance the ignition by means of the manual control. The grip on the paper should begin to be released just as the contact breaker cam reaches its full extent of travel.

6 To retime the ignition the magneto pinion will require removal from the magneto armature. A similar method as for checking must be adopted. Remember that the magneto rotates in a clockwise direction viewed from the contact breaker end.

7 On machines fitted with a timed breather, driven from a peg on the pinion, the magneto pinion must be aligned so that the drive peg is exactly in line with the timing cover screw hole immediately above the idler pinion when the piston is at TDC (see Fig. 1.8 in Chapter 1).

8 When timing the Gold Stars or the competition models, it is unlikely that the timing stick method will be sufficiently accurate. Under these circumstances, the use of a degree disc, attached to the end of the mainshaft, is advised. See Specifications for degree readings.

10 Contact breaker adjustment

Alternator models only

1 The contact breaker gap is correct when the blade of a 0.012 in (0.30 mm) feeler gauge is a light sliding fit between the points faces with the contact breaker in the fully open position.

2 To gain access to the contact breaker remove the cover, which

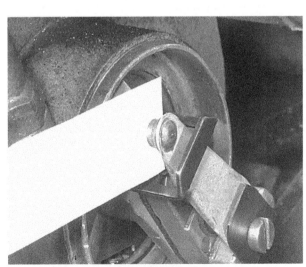

9.5 Use cigarette paper slip to ascertain points opening

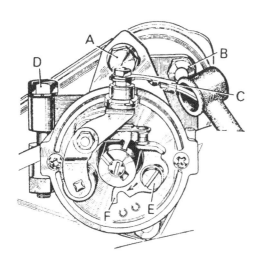

Fig. 5.2. The 1958-60 contact breaker assembly

A Adjuster bolt   D Clamp bolt
B Nut   E Points locking bolt
C Terminal   F Adjusting slot

is retained by two clips. Adjust the contact breaker points by
loosening the fixed point retaining screw and moving the point
with a screwdriver blade. Tighten the screw and recheck the
gap.

11 Contact breaker: removing, cleaning and replacement

1 If on inspection the points faces are found to be only
slightly soiled, they may be cleaned in-situ using a fine swiss
file or strip of emery paper (No. 400) backed by a thin strip of
tin. Badly blackened, pitted or burnt points should be removed
for dressing or renewal.
2 To remove the points, detach the low tension lead at the
contact breaker housing terminal and remove the nut and washer
which holds the condenser in place. Lift the condenser from
position. After removal of the fixed point retaining screw the
contact breaker assembly may be lifted clear. Before separating
the two points, note the position of any insulating washer either
on the moving point pivot post or at the low tension terminal.
All components must be refitted in their original positions to
ensure correct functioning of the contact breaker.
3 The points faces should be cleaned as described in Section
5.5. If a large amount of metal has to be removed in order to
restore the points faces, the contact breaker assembly should
be renewed.

12 Auto-advance unit

Alternator models only
1 An automatic advance-retard unit is incorporated in the
contact breaker housing assembly, to ensure that the optimum
ignition timing is given at all engine speeds.
2 Access to the auto-advance unit can be gained by removing
the contact breaker base plate, which is retained by two crosshead screws.
3 Provided that the auto-advance bob-weight pivots are
adequately lubricated, the assembly should give little trouble.
A few drops of oil to each pivot at regular intervals is sufficient.
Eventually the bob-weight control springs will weaken, altering
the advance characteristics. Only springs of the correct type
should be refitted.

13 Ignition timing: checking and resetting

Alternator models only
1 The ignition may be checked in a similar manner to that
given for magneto models in Section 9. When checking the
points for correct opening, the cam may be advanced manually
by turning in a clockwise direction to the full limit of travel.
A small amount of adjustment in ignition timing may be made
by slackening the bolt passing through the housing clamp
bracket and moving the bracket to the left or right, within the
limits of the elongated hole.
2 If the complete contact breaker housing has been removed,
or the cam drive pinion has been detached, ignition timing
should be carried out as follows:
    Position the piston at 7/16 in (11.0 mm) B31 or 3/8 in (9.5
mm) B33 BTDC on the compression stroke using the method
described in Section 9.3.
3 Fit the contact breaker drive pinion to the shaft and line up
the radial drilling in the pinion boss with that in the shaft. Insert
the drive pin and fit the securing clip. Place a new gasket on the
contact breaker housing. A few blobs of gasket compound on
the gasket will help retain it in position.
4 Check that the low tension terminal on the contact breaker
housing is in alignment with the bolt passing through the adjustment bracket and into the aluminium housing. If the terminal
does not align, loosen the slotted clamp bolt and turn the complete contact breaker base. Retighten the clamp bolt.
5 Turn the contact breaker cam in a clockwise direction,
independently of the driveshaft, until it is in the fully advanced
positon. Now turn the cam further in the same direction, until
the points are just on the verge of opening. Release the hold on
the cam. Ensuring that the relative positions between the drive
pinion and the housing do not alter, insert the complete unit
into position so that the drive pinion passes into mesh with the
idler pinion. The bracket bolt and low tension terminal must
be vertical whilst this is carried out.
6 If the drive pinion does not mesh with the idler pinion
turn it the fraction of a tooth necessary to ensure meshing.
Temporarily refit the three aluminium housing retaining screws
and nuts, and check the timing. Slight inaccuracy may be
rectified by loosening the bracket bolt and moving the bracket
within the limits of the elongated hole, through which the bolt
passes.

14 Condenser: faults

Alternator model
1 A condenser is incorporated in the contact breaker circuit to
prevent arcing across the points, which would quickly burn the
points and cause the ignition to fail.
2 Badly blackened or burnt points are caused by a breakdown
in the condenser. Repair of a condenser is not a practicable
proposition; a new component should be fitted. The condenser
is located adjacent to the contact breaker points and is retained
in position by a nut and washer, screwed onto the moving
contact pivot post.

15 Ignition coil: checking

1 The ignition coil is a sealed unit, designed to give long service.
It is normally attached by two bolts to the inside of the tool
box, where it is amply protected from the effects of damp and
vibration.
2 To test the coil, first ensure that the terminals are clean and
tight and the HT lead is correctly connected within the coil.
Check that the battery is fully charged and remove the contact
breaker dust cover, turn the engine over slowly until the points
are closed.

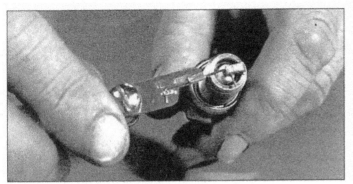

Electrode gap check - use a wire type gauge for best results

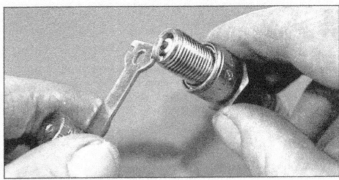

Electrode gap adjustment - bend the side electrode using the correct tool

Normal condition - A brown, tan or grey firing end indicates that the engine is in good condition and that the plug type is correct

Ash deposits - Light brown deposits encrusted on the electrodes and insulator, leading to misfire and hesitation. Caused by excessive amounts of oil in the combustion chamber or poor quality fuel/oil

Carbon fouling - Dry, black sooty deposits leading to misfire and weak spark. Caused by an over-rich fuel/air mixture, faulty choke operation or blocked air filter

Oil fouling - Wet oily deposits leading to misfire and weak spark. Caused by oil leakage past piston rings or valve guides (4-stroke engine), or excess lubricant (2-stroke engine)

Overheating - A blistered white insulator and glazed electrodes. Caused by ignition system fault, incorrect fuel, or cooling system fault

Worn plug - Worn electrodes will cause poor starting in damp or cold weather and will also waste fuel

3   Wedge the spark plug end of the HT lead between two
cylinder barrel fins so that the bared end of the centre wire is
approximately 3/16-¼ in (5-6 mm) away from the metal of
the cooling fins.

   Switch on the ignition and, using a plastic handled screwdriver flip the contact breaker points open - a healthy and
quite audible spark should jump from the end of the HT lead
to the cylinder barrel fins. Repeat the operation several times,
then switch off the ignition to avoid damage to the coil.

4   If no spark results and it is known that the battery, condenser and contact breaker points are not at fault, (see preceding
paragraphs) take it to an electrical repair expert for checking.
A faulty coil must be renewed as it is not practicable to effect
a repair.

16 Spark plugs: checking and resetting the gap

1   A 14 mm spark plug is fitted to all BSA single cylinder
models, irrespective of whether the cylinder head is cast in iron
or aluminium alloy. Refer to the Specifications Section for the
list of recommended grades.

2   Models fitted with a cast iron cylinder head use a spark
plug with a ½ in reach; ¾ in reach plugs are required for all
models fitted with an aluminium alloy cylinder head. Always
use the grade of plug recommended, or the direct equivalent in
another manufacturer's range.

3   Check the gap at the plug points every 2,000 miles. To reset
the gap, bend the outer electrode closer to the central electrode
and check that a 0.018 in feeler gauge can be inserted. Never
bend the central electrode, otherwise the insulator will crack,
causing engine damage if particles fall in whilst the engine is
running.

4   The condition of the spark plug electrodes and insulator
can be used as a reliable guide to engine operating conditions,
with some experience. See accompanying illustrations.

5   Always carry one spare spark plug of the correct grade. This
will serve as a get-you-home means if the spark plug in the engine
should fail.

6   Never overtighten a spark plug, otherwise there is risk of
stripping the threads from the cylinder head, especially in the
case of one cast in light alloy. A stripped thread can be repaired
by using what is known as a 'Helicoil' thread insert, a low cost
service of cylinder head reclamation that is operated by many
dealers.

7   Use a spark plug spanner that is a good fit, otherwise the
spanner may slip and break the insulator. The plug should be
tightened sufficiently to seat firmly on its sealing washer.

8   Make sure the plug insulating cap is a good fit and free from
cracks. The cap contains the suppressor that eliminates radio and
TV interference; in rare cases the suppressor has developed a
very high resistance as it has aged, cutting down the spark
intensity and giving rise to ignition problems.

17 Fault diagnosis: Ignition system

Symptom	Cause	Remedy
Engine will not start	No spark at plug	Check whether points open and close. Check also whether points are dirty - if so, clean. Check whether points arc when engine is turned over. If so, condenser has failed. Replace armature (magneto models). Replace condenser (alternator models). Cut-out earthed. Remove lead and check whether plug sparks.
Engine starts, but runs erratically	Intermittent or weak spark	Try renewing. Check ignition timing. Check plug lead for short circuits.
Engine will not run at low speeds, kicks back during starting	Ignition over-advanced	Re-set ignition timing.
Engine lacks power, overheats	Ignition timing retarded	Re-set ignition timing.
Engine misfires at high speeds	Incorrect spark plug	Check with list of recommendations.

Chapter 6 Frame and forks

Contents

General description	... ...	1
Front forks: removal from frame	... ...	2
Front forks: dismantling	... ...	3
Front forks: general examination	... ...	4
Front forks: examination and replacement of oil seals	...	5
Front forks: examination and replacement of bushes	...	6
Steering head bearings: examination and replacement	...	7
Front forks: reassembly	... ...	8
Front forks: damping action	... ...	9
Frame assembly: examination and renovation	... ...	10
Swinging arm rear suspension: examination and renovation	... ...	11
Rear suspension units: examination	... ...	12
Rear suspension units: adjusting the setting	... ...	13

Centre stand: examination	... ...	14
Prop stand: examination	... ...	15
Footrests: examination and renovation	... ...	16
Speedometer: removal and replacement	... ...	17
Speedometer cable: examination and replacement	...	18
Tachometer: removal and replacement	... ...	19
Tachometer drive cable: examination and renovation	...	20
Tachometer drive gearbox: examination	... ...	21
Dualseat: removal	... ...	22
Tank badges and motifs	... ...	23
Steering head lock	... ...	24
Steering damper: use	... ...	25
Cleaning: general	... ...	26
Fault diagnosis: frame and forks	... ...	27

Specifications

Front forks
All models   ... ...   Telescopic, external spring, one-way hydraulically damped

Frame
M series   ... ...   Spring frame (plunger type)
All others   ... ...   Swinging arm frame controlled by two hydraulically damped rear suspension units

Damping fluid
Oil content per fork leg   ... ...   213 cc (3/8 Imp pint)
Oil viscosity   ... ...   SAE 20 oil

1 General description

1 Three different frame types were used on various models within the range covered by this manual. The M series machines were available from 1954 with optional rigid or spring frames (plunger) which, with the exception of the M33 model, was standard up to the end of production. The spring frame only was available for the M33 model from 1956 on. On B series machines, including the B31, B32, and B33 and B34 models, swinging arm rear suspension was introduced in 1954 and was available as an optional extra until 1956, when the plunger suspension was phased out. All Gold Star models covered in this manual are fitted with swinging arm frames.
2 Despite the changes in frame specification, the telescopic front forks fitted to all models have been modified only superficially, the forks from one model to another being fundamentally the same.
3 The rear suspension units fitted to swinging arm models are hydraulically damped and have a three position cam adjustment,

to enable effective changes in the spring rate to be made to suit varying riding conditions.

2 Front forks: removal from frame

1 It is unlikely that the front forks will need to be removed from the frame as a complete unit, unless the steering head bearings require attention or the forks are damaged in an accident.
2 Commence operations by placing the machine on the centre stand or by placing a stout wooden box below the crankcase to raise the front wheel off the ground (early models without a centre stand). Disconnect the front brake cable by removing the nut and bolt through the brake operating arm and unscrewing the cable adjuster from the brake plate. The operating cable can now be pulled clear and if necessary, detached from the front brake lever. On some models it is necessary also to detach the front brake torque arm.

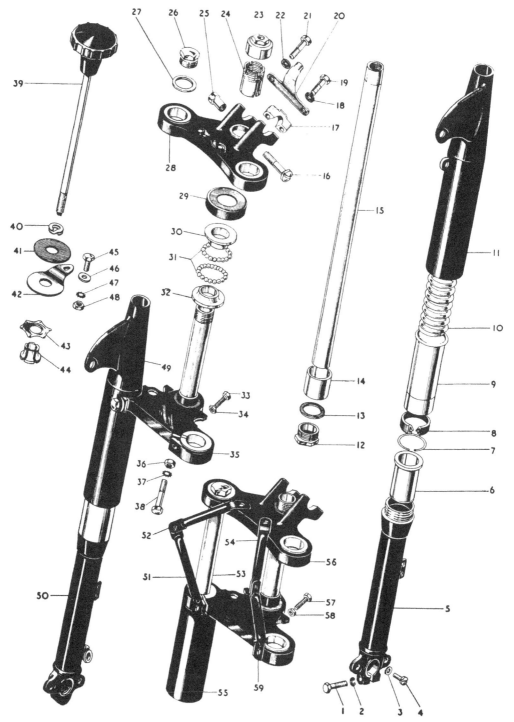

Fig. 6.1. Front forks - component parts, Gold Star models

1	Clamp bolt	16	Clamp bolt	31	Steering head balls (40 off)	46	Washer
2	Spring washer	17	Handlebar clamp (2 off)	32	Bottom cone	47	Washer
3	Fibre washer (2 off)	18	Washer (4 off)	33	Stop bolt (2 off)	48	Nut
4	Drain plug (2 off)	19	Bolt (2 off)	34	Lock nut (2 off)	49	Fork shroud
5	Lower fork leg	20	Spring plate	35	Bottom yoke	50	Lower fork leg
6	Top fork bush (2 off)	21	Bolt (2 off)	36	Nut (2 off)	51	Lower lamp bracket
7	Circlip (2 off)	22	Spacer (2 off)	37	Washer (2 off)	52	Upper lamp bracket
8	Oil seal (2 off)	23	Sleeve nut	38	Bolt (2 off)	53	Stanchion (inner fork leg)
9	Oil seal holder (2 off)	24	Adjusting sleeve	39	Damper rod		(2 off)
10	Fork spring (2 off)	25	Nut	40	Double spring washer	54	Upper lamp bracket
11	Fork shroud	26	Fork cap (2 off)	41	Friction washer	55	Lower fork shroud (2 off)
12	Stanchion plug (2 off)	27	Fibre washer (2 off)	42	Anchor plate	56	Top yoke
13	Washer (2 off)	28	Top yoke	43	Pressure washer	57	Bolt (2 off)
14	Bottom fork bush (2 off)	29	Dust cap	44	Damper centre	58	Nut (2 off)
15	Stanchion (inner fork leg)	30	Top cone	45	Bolt	59	Lower lamp bracket
	(2 off)						

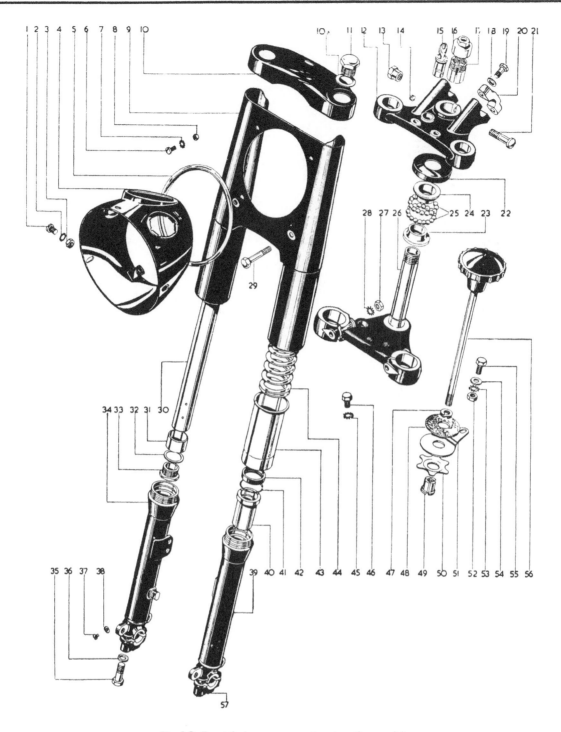

Fig. 6.2. Front forks - component parts, other models

1	Headlamp adjusting screw (2 off)	15	Lock assembly	30	Stanchion (inner fork leg) (2 off)	43	Oil seal holder (2 off)

1 Headlamp adjusting screw (2 off)
2 Washer (2 off)
3 Nut (2 off)
4 Headlamp cowl
5 Mounting ring
6 Screw (4 off)
7 Washer (4 off)
8 Nut (4 off)
9 Top cover
10 Yoke cover
10a Washer (2 off)
11 Nut (2 off)
12 Top fork yoke
13 Nut

15 Lock assembly
16 Sleeve cap
17 Sleeve
18 Washer (4 off)
19 Bolt (4 off)
20 Clamp (2 off)
21 Bolt
22 Dust cap
23 Bottom cone
24 Top cone
25 Steering head balls (40 off)
26 Bottom fork yoke
27 Nut (2 off)
28 Washer (2 off)
29 Bolt (2 off)

30 Stanchion (inner fork leg) (2 off)
31 Bush (2 off)
32 Washer (2 off)
33 Stanchion plug (2 off)
34 Lower fork leg
35 Bolt (4 off)
36 Spring washer (4 off)
37 Drain plug (2 off)
38 Fibre washer (2 off)
39 Lower fork leg
40 Stanchion upper bearing (2 off)
41 Circlip (2 off)
42 Oil seal (2 off)

43 Oil seal holder (2 off)
44 Fork spring (2 off)
45 Washer (2 off)
46 Bolt (2 off)
47 Washer
48 Friction washer (2 off)
49 Damper centre
50 Pressure plate
51 Anchor plate
52 Nut
53 Washer
54 Washer
55 Bolt
56 Steering damper rod
57 Fork end cap (2 off)

2.3 Remove spindle clamps to detach front wheel (late models)

2.7a Unscrew the chrome top caps and ...

2.7b ... slacken the fork leg pinch bolts

3   Remove the two bolts from the extreme end of both fork legs so that the bottom half of the split clamp arrangement can be detached. The front wheel will now be released from the fork ends, complete with brake plate and spindle. It may be necessary to turn the forks at an angle so that the front wheel can be lifted clear, away from the mudguard. Fit a block under the centre stand to gain sufficient clearance. Early models have a spindle and nut arrangement. The spindle is withdrawn after the nut is removed and the pinch bolt on the left-hand fork leg is slackened.

4   Detach the controls from the handlebars by either disconnecting the cable ends or by detaching the controls with the cables still attached. This includes the cut-out button, horn button and dip switch. In the case of the horn button, it is preferable to detach the main lead from the battery before the button is removed, to prevent short circuits. Detach the speedometer drive cable by unscrewing the gland nut from the bottom of the speedometer head and the pilot bulb within the speedometer head.

5   Remove the handlebars by withdrawing the bolts that retain the split mounting clamps to the fork top yoke.

6   Remove the headlamp by unscrewing the two retaining bolts in each side of the shell. If a cowl is fitted, this must be removed with the headlamp. The headlamp can be left to hang in a position where it is not liable to suffer damage. On late models, the headlamp nacelle is integral with the fork shrouds, in which case the front of the headlamp should be removed, the speedometer drive cable disconnected and the wires detached from the headlamp switch. It is advisable to make a note of the connections, even if the colour coding of the wires aids correct replacement.

7   Unscrew and remove the two chromium plated caps at the top of each fork leg. Slacken the pinch bolt in the fork top yoke, immediately to the rear of the steering head and remove the chromium plated cap at the top of the steering head. This will expose the slotted adjusting sleeve, which should be unscrewed (right-hand thread) using either a strip of metal to engage with the slots or the BSA Service Tool designed for this purpose. If the machine is fitted with a steering damper, it will be necessary first to remove the split pin from the bottom end of the stem and unscrew the knob until it can be detached, complete with rod.

8   When the adjusting sleeve has been withdrawn, the fork top yoke can be removed by striking it from the underside with a rawhide mallet, first one side and then the other. The forks, complete with steering head stem can now be drawn downwards until they are completely separate from the machine. Note that the uncaged ball bearings of the steering head assembly will drop free as the cups and cones separate, necessitating some arrangement for catching them.

9   If further dismantling is necessary, the front mudguard can be removed after the forks have been separated from the frame by disconnecting the mudguard stays and the bridge piece across the fork legs.

3   Front forks: dismantling

1   To remove the individual fork legs, unscrew the pinch bolts in the fork bottom yoke, and unscrew the clips securing the rubber gaiters (where fitted). The fork legs should now pull clear. If they are still a tight fit, spring open each pinch bolt a little, to ease the grip.

2   At this stage it is preferable to drain each fork leg by removing the drain plug in the lower extremity, close to the point through which the wheel spindle passes. Pull off the fork springs.

3   Unscrew the chromium plated cover immediately above the lower fork leg. This has a right-hand thread and may require the use of a strap spanner or BSA Service Tool No. 61—3005 to unscrew it. When the cover has been removed, it will expose a circlip within the lower fork leg that acts as a limit stop. If the circlip is removed, the inner fork leg can be withdrawn

2.9 Detach front mudguard

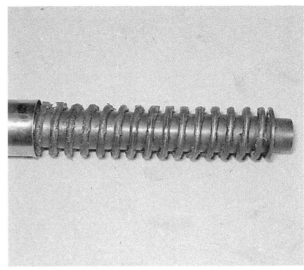

3.2 Remove the fork spring

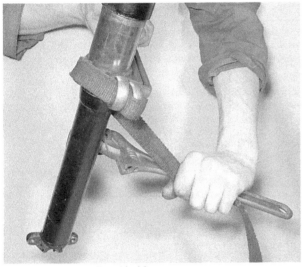

3.3a Unscrew fork oil seal holder

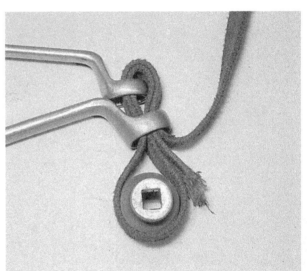

3.3b Example of fabricated strap wrench

3.3c Lift oil seal holder off stanchion

3.3d Prise upper slider retaining circlip from groove and ...

3.3e ... remove together with backing washer

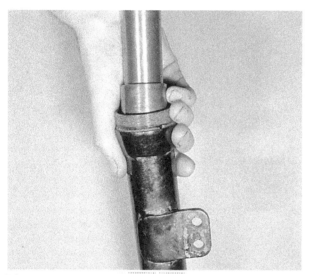

3.4 Withdraw the fork stanchion from lower leg

6.1a Nut and washer secure lower bush

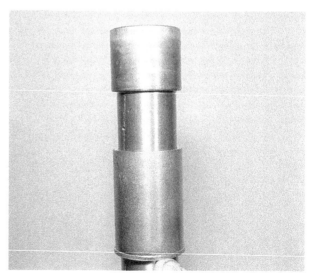

6.1b Check fit of bushes on stanchion

completely, together with the fork bushes.

4   It is not necessary to disturb the head race assembly or to detach the fork yokes from the frame if the fork legs alone are to be dismantled. The legs can be dismantled individually, as described, if the front wheel and mudguard are removed.

4   Front forks: general examination

1   Apart from the oil seals and bushes, it is unlikely that the forks will require any additional attention, unless the fork springs are weak or have to be replaced with stronger springs when the machine is to be used with a sidecar attached. If the fork legs or yokes have been damaged in an accident, it is preferable to have them replaced. Repairs are seldom practicable without the appropriate repair equipment and jigs, furthermore there is also the risk of fatigue failure.

2   Visual examination will show whether either the fork legs or the yokes are bent or distorted. The best check for the fork legs is to remove the fork bushes, as described in Section 6 of this Chapter and roll the legs on a sheet of plate glass. Any deviation from parallel will immediately be obvious.

5   Front forks: examination and replacement of oil seals

1   If the fork legs have shown a tendency to leak oil or if there is any other reason to suspect the condition of the oil seals, now is the time to replace them.

2   The oil seals are retained within the plated covers that thread on to the bottom legs of the forks. A BSA Service Tool No. 61—3006 is available for extracting the oil seals but since the seals have to be replaced there is no reason why they should not be drifted out of position through the slots in the plated covers. It follows that the seals should not be disturbed unless replacement is necessary.

6   Front forks: examination and replacement of bushes

1   Some indication of the extent of wear of the fork bushes can be gained when the forks are being dismantled. Pull each fork inner tube out until it reaches the limit of its extension and check the side play. In this position the two fork bushes are closest together, which will show the amount of play to its maximum. Only a small amount of play that is just perceptible

can be tolerated. If the play is greater than this, the bushes are due for replacement.

2 It is possible to check for play in the bushes whilst the forks are still attached to the machine. If the front wheel is gripped between the knees and the handlebars rocked to and fro, the amount of wear will be magnified by the leverage at the handlebar ends. Cross-check by applying the front brake and pushing and pulling the front wheel backwards and forwards. It is important not to confuse any play that is evident with slackness in the steering head bearings, which should be taken up first.

3 The fork bushes can be slid off the fork tubes if the tubes are clamped in a vice fitted with soft clamps and the large nut on the extreme end removed. If the replacement bushes are a slack fit on the tubes, wear has occurred on the tubes also, in which case a specialist repair with undersize bushes must be made.

4 The fit within the lower fork legs is also important. If wear of the inner surface is evident, it may be necessary to fit lower bushes that have a slightly greater outside diameter.

7 Steering head bearings: examination and replacement

1 Before commencing to reassemble the forks, inspect the steering head races. The ball bearing tracks should be polished and free from indentations and cracks. If signs of wear are evident, the cones and cups must be replaced. They are a tight press fit and must be drifted out of position. A BSA Service Tool No. 61-3063 is available for extracting the cups that remain within the steering head assembly of the frame. It screws into the threaded centre of each cup and is driven out from the opposite end, bringing the cup with it.

2 Ball bearings are cheap. If there is any reason to suspect the condition of the existing ball bearings, they should be replaced without question. Note that each race is not completely full of ball bearings. Space should be left for the theoretical insertion of one extra ball, so that the race is not crowded, forcing the ball bearings to skid against one another.

3 Use thick grease to retain the ball bearings in position, whilst the head stem is being assembled and adjusted.

8 Front forks: reassembly

1 To reassemble the forks, follow the dismantling procedure in reverse. Take particular care when passing the sliding fork members through the oil seals, which should be fitted with the lip facing downwards. It is a wise precaution to wind a turn or so of medium twine around the undercut at the base of the thread of the plated collars, to act as an extra seal.

2 Tighten the steering head carefully, so that all play is eliminated without placing undue stress on the bearings. The adjustment is correct if all play is eliminated and the handlebars will swing to full lock of their own accord when given a light push on one end.

3 It is possible to place several tons pressure quite unwittingly on the steering head bearings, if they are over-tightened. The usual symptom of over-tight bearings is a tendency for the machine to roll at low speeds, even though the handlebars may appear to turn quite freely.

4 One problem that will arise during reassembly is the reluctance of the fork main tubes to pass up into the fork top yoke. BSA Service Tool No. 61—3350 is used for this purpose; it threads into the top of each fork tube and can be used to pull the tube upward so that the tapered end engages with the fork yoke. If the tool is not available, a broom handle of the correct diameter can be used with equal effect, if it is first screwed into the end of the thread of each fork tube. Care should be taken in this instance, to prevent particles of wood from falling into the fork tubes.

5 If, after assembly, it is found that the forks are incorrectly aligned or unduly stiff in action, loosen the front wheel spindle, the two caps at the top of the fork legs and the pinch bolts in both the top and bottom yokes. The forks should then be

6.2 Oil seal can be prised from place

8.4a Pass broom handle through fork shroud and ...

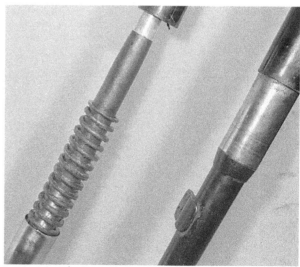
8.4b ... into leg when refitting

pumped up and down several times to realign them. Retighten all the nuts and bolts in the same order, finishing with the steering head pinch bolt.

6   This same procedure can be used if the forks are misaligned after an accident. Often the legs will twist within the fork yokes, giving the impression of more serious damage, even though no structural damage has occurred.

7   Do not omit to add the correct amount of damping oil to each fork leg before replacing the fork leg caps. See Specifications list for the amount and viscosity of oil to be added.

9   Front forks: damping action

1   Each fork leg contains a predetermined quantity of oil of recommended viscosity, which is used as a damping medium to control the action of the compression springs within the forks when various road shocks are encountered. If the damping fluid is absent, there is no control over the rebound action of the fork springs and fork movement will be excessive, giving a very 'lively' ride. Damping restricts fork movement on the rebound and is progressive in action — the effect becomes more powerful as the rate of deflection increases.

2   In the BSA system, the oil is contained in the lower fork leg. When the forks are deflected, the space between the upper and lower fork bushes becomes greater and oil enters the inner fork tube, via the large diameter hole in the nut at the bottom of each fork tube, under force. Small diameter holes in the inner tube allow the oil to enter the space between the two fork bushes, but they have a restrictive effect, which slows down the movement of the sliding tube. The rate of damping is governed by the size of these holes — the smaller the holes, the greater the damping effect. Before the sliding tube can reach its limit of travel, a tapered plug in the bottom of the lower fork leg enters the hole in the nut of the inner fork tube and slows down the rate of oil transfer until it is virtually cut off altogether. The remaining oil is incompressible and the fork leg is therefore prevented from 'bottoming' in a most effective manner. When the fork leg moves in the opposite direction (rebound) the space between the two fork bushes is reduced and the oil contained within this space must return through the holes drilled in the inner fork tube. As the fork movement increases still further, the uppermost bush will cover the holes completely, thereby leaving a quantity of oil in the space that cannot be compressed. The rebound action ceases as a result — it has been damped out.

3   The damping action can be varied only by changing the viscosity of the oil used as the damping medium; it is not practicable to vary the size of the holes in the inner fork tubes. In temperate climates an SAE 20 oil is used but if considered necessary, the viscosity rating can be increased without any harmful effects.

10 Frame assembly: examination and renovation

1   A rigid frame was used for some M series models which should not require attention unless it is damaged in an accident. Frame repairs are best entrusted to a specialist in this type of repair work, who will have all the necessary jigs and mandrels available to ensure correct alignment. In many instances a replacement frame from a breaker's yard is the cheaper and more satisfactory alternative.

2   If the machine is stripped for an overhaul, this affords an excellent opportunity to inspect the frame for signs of cracks or other damage that may have occurred in service. Check the front down tube at the point immediately below the steering head, which is where a break is most likely to occur. Check the top tube of the frame for straightness — this is the tube most likely to bend in the event of an accident.

3   Many M and B series models employ a plunger form of rear suspension, in which the vertical movement of the rear wheel is controlled by undamped coil springs. Wear is likely to occur in the bushes of the sliding member and it is necessary to dismantle each unit, in order to gain access.

4   Remove the rear wheel as detailed in Chapter 7, Section 6. Disconnect the mudguard stays and detach both silencers. Withdraw the pinch bolts from the top and bottom of each suspension unit.

5   Take out the plug that threads into the top of each centre column and insert a metal rod tapped with the same thread or BSA Service Tool No. 61-3222. Use this to tap the column out of position so that it is withdrawn from the underside.

6   The remainder of each unit can now be pulled from between the upper and lower lugs of the frame end, an operation made easier if the unit is compressed a little. The springs will be released when the units emerge from the fork ends, but they are not under any great compression.

7   Remove the shrouds and springs from the sliding members, taking note of the position of the various components, to aid replacement. The bushes in the sliding member can now be examined for wear; should they require replacing the tube complete with bushes must be replaced. To withdraw the tube, remove the centre pinch bolt and if necessary, open the slot with the blade of a screwdriver.

8   Replace the units by following the dismantling procedure in reverse. When replacing the sliding member, note that the hole in the side must line up with the grease nipple in the fork end. Place the lower shroud in the bottom frame lug and press down on the upper shroud to compress the springs. Make sure the units are replaced in the correct order, because the nearside fork end carries the anchor lug for the rear brake plate.

9   Some 1954 machines and all those except the M series made after 1954, employ swinging arm rear suspension in which the rear sub-frame pivots from a point to the rear of the gearbox. Movement is controlled by two hydraulically-damped rear suspension units. More detailed attention is required when this form of suspension requires inspection and it is necessary to devote the following complete Section to this quite differer . mode of rear suspension.

11 Swinging arm rear suspension: examination and renovation

1   After an extended period of service, the bush and pivot pin of the swinging arm fork will wear, giving rise to lateral play that will affect the handling characteristics of the machine. The bushes are of the Silentbloc type, held in frame lugs close to the gearbox centre. They are a tight press fit and cannot be removed

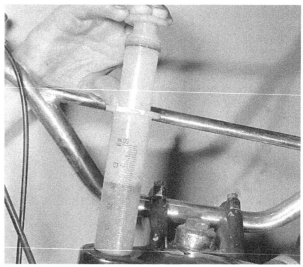

8.7 Do not omit damping fluid before replacing caps

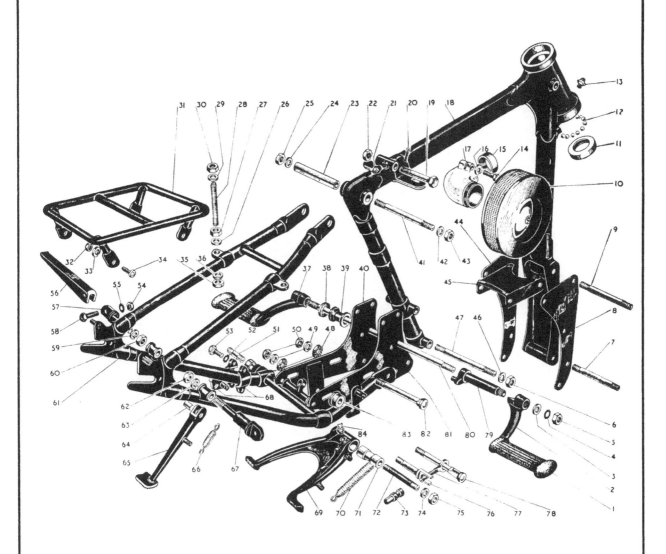

Fig. 6.3. Rigid frame - component parts

1	Footrest rubber - 2 off	22	Nut	43	Nut	64	Pivot pin
2	Footrest	23	Sleeve	44	Engine plate	65	Prop stand
3	Washer - 2 off	24	Washer	45	Engine plate cover	66	Spring
4	Washer - 2 off	25	Nut	46	Spring washer - 19 off	67	Bolt - 2 off
5	Nut - 2 off	26	Washer - 2 or 4 off	47	Stud	68	Prop stand lug
6	Nut - 19 off	27	Nut - 2 off	48	Serrated washer - 2 off	69	Centre stand
7	Engine plate stud - 7 off	28	Stud - 2 off	49	Spring washer - 2 off	70	Centre stand spring
8	Engine plate	29	Washer - 2 off	50	Nut - 2 off	71	Distance piece
9	Downtube stud	30	Nut - 2 off	51	Stud - 1 or 2 off	72	Stud
10	Air cleaner	31	Carrier	52	Washer - 2 off	73	Adjuster body
11	Bearing cap - 2 off	32	Carrier	53	Bolt - 2 off	74	Spring washer - 2 off
12	Steering head balls - 40 off	33	Washer - 4 off	54	Nut - 2 off	75	Nut - 2 off
13	Grease nipple	34	Screw - 4 off	55	Washer - 2 off	76	Stop
14	Bolt	35	Nut - 2 off	56	Pillion footrest - 2 off	77	Adjuster bolt
15	Adaptor	36	Washer - 2 off	57	Bolt	78	Adjuster mounting bolt
16	Washer	37	Footrest	58	Bolt - 2 off	79	Footrest splined mounting
17	Elbow	38	Cork washer - 2 off	59	Washer	80	Footrest rod
18	Front frame	39	Distance piece - 2 off	60	Nut	81	Rear engine/gearbox plate
19	Saddle bolt	40	Rear engine/gearbox plate	61	Rear frame section	82	Gearbox connecting bolt
20	Split pin	41	Stud	62	Nut	83	Spacer
21	Grease nipple	42	Washer	63	Washer	84	Grease nipple

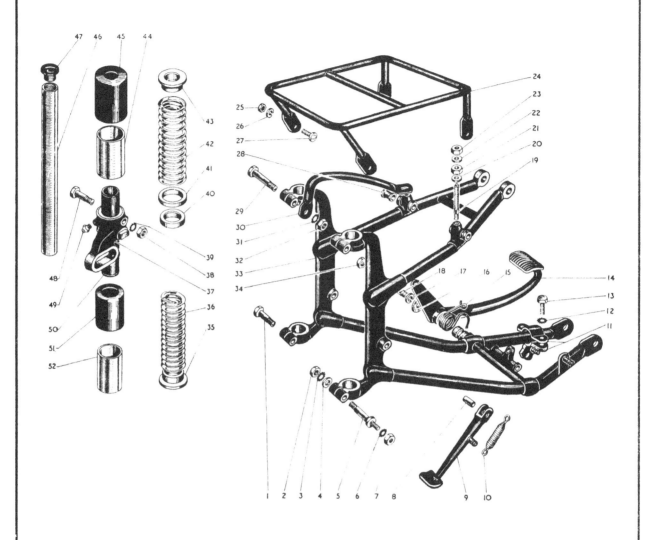

Fig. 6.4. Rear frame, plunger models - component parts*

1	Plunger clamp bolt	15	Return spring	29	Plunger clamp bolt - 2 off	41	Compression spring
2	Nut - 4 off	16	Washer	30	Lifting handle		bottom collar - 2 off
3	Washer - 4 off	17	Spring washer	31	Washer - 4 off	42	Compression spring - 2 off
4	Spacer - 2 off	18	Stud	32	Nut - 4 off	43	Compression spring top
5	Plunger clamp spindle	19	Stud - 2 off	33	Rear frame		collar - 2 off
6	Washer	20	Washer - 2 off	34	Nut	44	Inner shroud - 2 off
7	Nut	21	Nut - 2 off	35	Rbound spring collar - 2 off	45	Outer shroud - 2 off
8	Pivot pin	22	Washer - 2 off	36	Rebound spring - 2 off	46	Plunger column - 2 off
9	Prop stand	23	Nut - 2 off	37	Rear spindle lug	47	Top plug - 2 off
10	Prop stand spring	24	Carrier	38	Nut - 2 off	48	Lug bolt - 2 off
11	Prop stand lug	25	Nut - 4 off	39	Washer - 2 off	49	Grease nipple
12	Washer - 2 off	26	Spring washer - 4 off	40	Rebound spring top	50	Plunger tube - 2 off
13	Bolt - 2 off	27	Screw - 4 off		collar - 2 off	51	Lower outer shroud - 2 off
14	Rear brake pedal	28	Distance piece			52	Lower inner shroud - 2 off

* used in conjunction with front half of rigid frame (see Fig. 6.3)

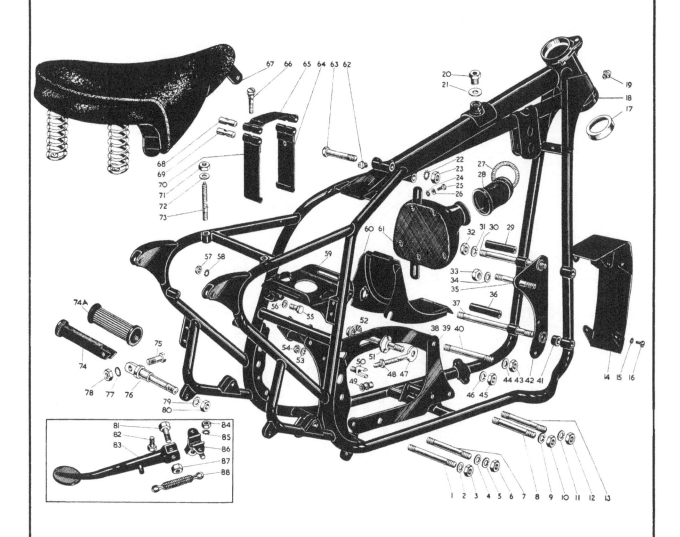

Fig. 6.5. Swinging arm frame - component parts

1 Gearbox mounting stud - 2 off	23 Nut	46 Spring washer	69 Tapped pivot
2 Spring washer - 5 off	24 Washer - 2 off	47 Gearbox adjuster	70 Rear battery strap
3 Nut - 5 off	25 Screw - 2 off	48 Locknut	71 Nut - 4 off
4 Washer - 2 off	26 Washer - 2 off	49 Adjuster nut	72 Washer - 4 off
5 Spring washer - 2 off	27 Clip	50 Stop	73 Saddle stud - 2 off
6 Nut - 2 off	28 Air cleaner hose	51 Serrated washer	74 Pillion footrest - 2 off
7 Gearbox pivot stud	29 Distance piece	52 Locking stud	74a Pillion footrest rubber - 2 off
8 Gearbox plate stud	30 Frame stud	53 Spring washer	75 Swivel bolt - 2 off
9 Spring washer	31 Spring washer - 4 off	54 Nut	76 Anchor bolt - 2 off
10 Nut - 2 off	32 Nut - 4 off	55 Bolt - 4 off	77 Washer - 2 off
11 Spring washer - 2 off	33 Nut - 6 off	56 Washer - 2 off	78 Nut - 2 off
12 Nut - 2 off	34 Spring washer 6 off	57 Nut -2 off	79 Washer - 2 off
13 Crankcase mounting stud	35 Crankcase stud - 3 off	58 Washer	80 Nut - 2 off
14 Engine plate cover	36 Distance piece	59 Battery base plate	81 Bolt
15 Washer - 6 off	37 Frame stud	60 Rear engine plate cover	82 Bolt - 2 off
16 Screw - 6 off	38 Gearbox top plate	61 Air cleaner	83 Prop stand
17 Bearing cup - 2 off	39 Rear engine plate - 2 off	62 Grease nipple	84 Nut
18 Frame	40 Crankcase stud	63 Bolt	85 Washer - 2 off
19 Grease nipple	41 Distance collar - 2 off	64 Front battery strap	86 Lug
20 Centre bolt boss	42 Front engine plate - 2 off	65 Top battery strap	87 Nut
21 Washer	43 Nut - 2 off	66 Bolt	88 Spring
22 Washer	44 Spring washer - 2 off	67 Saddle	
	45 Nut	68 Plain pivot	

11.1a Rear dampers are retained by bolts through lugs

11.1b Swinging arm pivot is retained by a nut and ...

11.1c ... by a small bolt passing through frame

without risk of damage unless the correct equipment is available. This form of repair is best entrusted to a BSA repair specialist who will have the appropriate equipment. It is highly improbable that the average rider/owner will have access to this equipment or the skill with which to undertake the reconditioning work necessary.

2   The rear suspension units are removed by withdrawing the upper and lower bolts, nuts, and washers.

12 Rear suspension units: examination

1   Only a limited amount of dismantling can be undertaken because the damper unit is sealed and cannot be dismantled. If the unit leaks oil, or if the damping action is lost, the unit must be replaced as a whole after removing the compression spring and shroud.

2   Before the spring and shroud can be removed, the unit must be detached from the machine and clamped in a vice. If pressure is applied to the top of the shroud, compressing the internal spring, the split collets can be removed and the spring and shroud released. Note the spring is colour-coded; the colour relates to the spring rating. Springs can be obtained in a variety of different ratings, to accommodate different loadings.

13 Rear suspension units: adjusting the setting

1   The Girling rear suspension units fitted to the BSA single cylinder models have a three-position cam adjuster built into the lower portion of the leg to suit varying load conditions. The lowest position should suit the average rider, under normal road conditions. When a pillion passenger is carried, the second or middle position offers a better choice and for continuous high speed work or off-the-road riding, the highest position is recommended.

2   These adjustments can be effected without need to detach the units. A 'C' spanner in the tool kit is used to rotate the cam ring until the desired setting is obtained.

14 Centre stand: examination

1   All models, excluding those fitted with a rigid frame, are provided with a centre stand attached to lugs on the bottom frame tubes. The stand provides a convenient means of parking the machine on level ground, or for raising one or other of the wheels clear of the ground in the event of a puncture. The stand pivots on a long bolt that passes through the lugs and is secured by a nut and washer. A return spring retracts the stand so that when the machine is pushed forward it will spring up and permit the machine to be wheeled, prior to riding.

2   The condition of the return spring and the return action should be checked frequently, also the security of the nut and bolt. If the stand drops whilst the machine is in motion, it may catch in some obstacle in the road and unseat the rider.

3   Rigid frame models have a rear stand pivoted on the extreme end of the frame. The stand is returned to the retracted position by an extension spring on the right-hand side of the machine.

15 Prop stand: examination

1   A prop stand that pivots from a lug at the front end of the lower left-hand frame tube provides an additional means of parking the machine. This too has a return spring, which should be strong enough to cause the stand to retract immediately the machine is raised into a vertical position. It is important that this spring is examined at regular intervals, also the nut and bolt that act as the pivot. A falling prop stand can have far more serious consequences if it should fall whilst the machine is on the move.

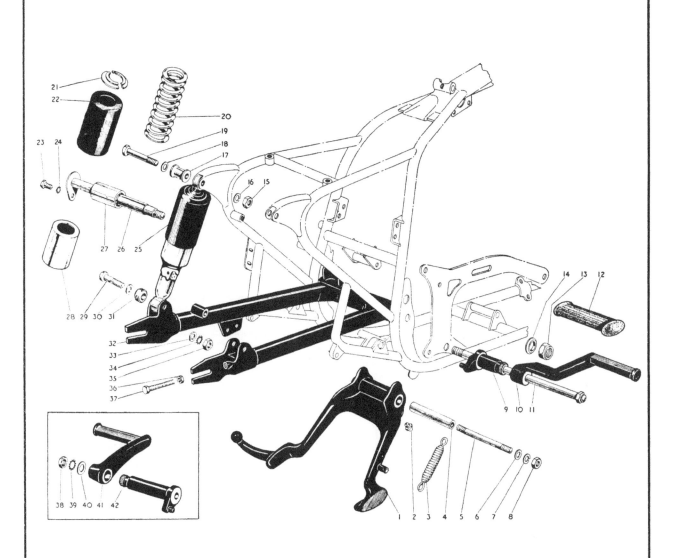

Fig. 6.6. Swinging arm rear suspension and stand - component parts

1 Centre stud	12 Footrest rubber - 2 off	23 Bolt	33 Washer - 2 off
2 Grease nipple	13 Locknut	24 Washer	34 Washer - 2 off
3 Spring	14 Washer	25 Damper assembly - 2 off	35 Nut - 2 off
4 Tube	15 Nut - 2 off	26 Swinging arm spindle	36 Locknut - 2 off
5 Pivot stud	16 Spring washer - 2 off	27 'Silentbloc' bush - 2 off	37 Adjusting screw - 2 off
6 Washer - 2 off	17 Spacer - 2 off	28 Inner shroud - 2 off	38 Nut
7 Spring washer - 2 off	18 Washer - 2 off	29 Mounting bolt - 2 off	39 Washer
8 Nut - 2 off	19 Bolt	30 Washer - 2 off	40 Washer
9 Footrest splined mounting	20 Damper spring - 2 off	31 Rubber bush - 4 off	41 Footrest
10 Footrest	21 Split shroud retainer	32 Swinging arm	42 Footrest splined mounting
11 Footrest stud	22 Damper shroud - 2 off		

16 Footrests: examination and renovation

1   The footrests, which bolt to the frame lugs, are malleable and will bend if the machine is dropped. Before they can be straightened, they must be detached from the frame and have the rubbers removed.
2   To straighten the footrests, clamp them in a vice and apply leverage from a long tube that slips over the end. The area in which the bend has occurred should be heated to a cherry red with a blow lamp, during the bending operation. Do not bend the footrests cold, otherwise there is risk of a sudden fracture.

17 Speedometer: removal and replacement

1   The BSA single cylinder models are fitted with a Smiths chronometric speedometer, calibrated up to 80 or 120 mph. An internal lamp is provided for illuminating the dial during the hours of darkness and the odometer has a trip setting, so that the lower mileage reading can be set to zero before a run is commenced.
2   The speedometer head has two studs, which permit it to be attached to either a bracket on the fork top yoke or to a nacelle or cowl by means of a simple clamp arrangement. When a clamp is fitted, removal of the retaining nut(s) will permit the speedometer head to be withdrawn from the top of the mounting, after the drive cable has been detached.
3   Apart from defects in the drive or the drive cable itself, a speedometer that malfunctions is difficult to repair. Fit a replacement or alternatively entrust the repair to an instrument repair specialist, bearing in mind that the speedometer must function in a satisfactory manner to meet Statutory requirements.
4   If the odometer readings continue to show an increase, without the speedometer indicating the road speed, it can be assumed the drive and drive cable are working correctly and that the speedometer head itself is at fault.

18 Speedometer cable: examination and renovation

1   It is advisable to detach the speedometer drive cable from time to time in order to check whether it is adequately lubricated, and whether the outer covering is compressed or damaged at any point along its run. A jerky or sluggish speedometer movement can often be attributed to a cable fault.
2   To grease the cable, withdraw the inner cable. After removing the old grease, clean with a petrol soaked rag and examine the cable for broken strands or other damage.
3   Regrease the cable with high melting point grease and ensure that there is no grease on the last six inches, at the end where the cable enters the speedometer head. If this precaution is not observed, grease will work into the speedometer head and immobilise the movement.
4   Inspection will show whether the speedometer drive cable has broken. If so, the inner cable can be removed and replaced with another whilst leaving the outer cable in place - provided the outer cable is not damaged or compressed at any point along its run. Measure the cable length exactly when purchasing a replacement, because this measurement is critical.

19 Tachometer: removal and replacement

1   The tachometer drive is normally taken from one of the timing pinions, using a modified timing cover that will accept the drive gearbox. The tachometer head may be of either the chronometric or magnetic type, depending on the year of manufacture.
2   It is not possible to effect a satisfactory repair to a defective tachometer head, hence replacement is necessary if the existing head malfunctions. Make sure an exact replacement is obtained;

some tachometer heads work at half-speed if a different type of drive gearbox is employed.
3   The tachometer head is illuminated internally so that the dial can be read during the hours of darkness.

20 Tachometer drive cable: examination and renovation

1   Although a little shorter in length, the tachometer drive cable is identical in construction to that used for the speedometer drive. The advice given in Section 18 of this Chapter applies also to the tachometer drive cable.

21 Tachometer drive gearbox: examination

1   The tachometer drive gearbox is unlikely to give trouble during the normal service life of the machine, provided it is greased regularly at the external grease nipple.
2   If the drive gearbox has to be replaced, make sure it is replaced with one having an identical drive ratio, otherwise the tachometer head readings will no longer be true.
3   There should be a good joint between the flange of the drive gearbox and the flange of the timing cover if oil leaks are to be prevented. Always use a new gasket and preferably a little gasket cement if the joint has to be broken and re-made.

22 Dualseat: removal

1   The dualseat is attached by means of two bolts towards the rear end of the underpan, one on each side of the machine. If these bolts are removed, the dualseat will lift off, after the slotted end of the nose connection has been pulled clear from the mounting tube across the frame.
2   Early models are fitted with a saddle that has a three-point fixing, a bolt through the nose of the saddle and a nut and bolt through each spring at the point where they joint the frame lugs. It is usually necessary to disconnect the nose fixing if the petrol tank has to be removed, so that the nose of the saddle can be tilted to provide sufficient clearance.

23 Tank badges and motifs

1   With the exception of some early models, which had various designs of transfer, all models are fitted with circular plastic tank badges retained on spring clips by two screws. The spring clips fit into a retaining clip welded to the petrol tank.
2   The colour of the badges and the style of motif displayed differs from model to model, and as such it is important to specify the model and colour scheme when endeavouring to obtain a replacement.

24 Steering head lock

1   Some models are fitted with a steering head lock inserted into the fork top yoke. If the forks are turned to the extreme left, they can be locked in this position to prevent theft. The lock is of Yale manufacture.
2   Add an occasional few drops of thin machine oil to keep the lock in good working order. This should be added to the periphery of the moving drum and NOT the keyhole.

25 Steering damper: use

1   Mention has been made of the steering damper, a friction plate device that can be adjusted to vary the amount of effort required to turn the handlebars. In many respects, the steering damper can be regarded as a legacy of the past when it was

necessary to counteract the tendency of some machines to develop a speed wobble at high speeds. Today, the steering damper comes into its own mainly when a sidecar is attached, since it will prevent the handlebars from oscillating at very low speeds, when it is applied.

2 Under normal riding conditions, the steering damper can be slackened off. It is sometimes advantageous to have it biting just a trifle at high speeds, to ease the strain on the arms.

3 To remove the steering damper, detach the split pin at the extreme end of the damper rod and/or unscrew the knob at the top of the steering head until it can be drawn away with the rod attached. When the fixed plate is detached from below the bottom yoke of the forks, the friction plates can be removed for inspection. They seldom require attention.

4 A steering damper is not necessarily fitted to all the models covered by this manual.

26 Cleaning: general

1 After removing all surface dirt with a rag or sponge that is washed frequently in clean water, the application of car polish or wax will restore a good finish to the cycle parts of the machine after they have dried thoroughly. The plated parts should require only a wipe with a damp rag, although it is permissible to use a chrome cleaner if the plated surfaces are badly tarnished.

2 Oil and grease, particularly when they are caked on, are best removed with a proprietary cleanser such as 'Gunk' or 'Jizer'. A few minutes should be allowed for the cleanser to penetrate the film of oil and grease before the parts concerned are hosed down. Take care to protect the magneto, carburettor(s) and electrical parts from the water, which may otherwise cause them to malfunction.

3 Polished aluminium alloy surfaces can be restored by the application of Solvol 'Autosol' or some similar polishing compound, and the use of a clean duster to give the final polish.

4 If possible, the machine should be wiped over immediately after it has been used in the wet, so that it is not garaged under damp conditions that will promote rusting. Make sure to wipe the chain and if necessary re-oil it, to prevent water from entering the rollers and causing harshness with an accompanying high rate of wear. Remember there is little chance of water entering the control cables if they are lubricated regularly, as recommended in the Routine Maintenance Section.

27 Fault diagnosis: frame and forks

Symptom	Cause	Remedy
Machine is unduly sensitive to road conditions	Forks and/or rear suspension units have defective damping	Check oil level in forks. Renew rear suspension units.
Machine tends to roll at low speeds	Steering head bearings overtight or damaged	Slacken bearing adjustment. If no improvement, dismantle and inspect bearings.
Machine tends to wander, steering is imprecise	Worn swinging arm bearings or sliders in plunger sprung models	Check and if necessary renew bearings.
Fork action stiff	Fork legs have twisted in yokes or have been drawn together at lower ends	Slacken off spindle nut clamps, pinch bolts in fork yokes and fork top nuts. Pump forks several times before retightening from bottom. Is distance piece missing from fork spindle?
Forks judder when front brake is applied	Worn fork bushes Steering head bearings too slack	Strip forks and replace bushes. Readjust, to take up play.
Wheels out of alignment	Frame distorted as result of accident damage	Check frame alignment after stripping out. If bent, specialist repair is necessary.

Chapter 7 Wheels, brakes and tyres

Contents

General description 1	Rear wheel bearings: examination and replacement ... 9
Front wheel: examination and renovation 2	Front and rear brakes: adjustment 10
Front brake assembly: examination, renovation and	Rear wheel sprocket: removal, examination and
reassembly 3	replacement 11
Wheel bearings: examination and replacement 4	Final drive chain: examination, lubrication and
Front wheel: reassembly and replacement 5	adjustment 12
Rear wheel: examination, removal and renovation 6	Rear wheel: replacement 13
Brake drum: removal, quickly-detachable steel hub wheel	Wheel balance 14
only 7	Tyres: removal and replacement 15
Rear brake assembly: examination, renovation and	Tyre valve dust caps 16
reassembly 8	Fault diagnosis 17

Specifications

Tyre sizes

	Front	Rear
M20	3.25 x 19 inch	3.25 x 19 inch
M21	3.50 x 19 inch	3.50 x 19 inch
M33	3.25 x 19 inch	3.50 x 19 inch
B31	3.25 x 19 inch	3.25 x 19 inch
B33	3.25 x 19 inch	3.50 x 19 inch
B32 and B34 ... ... competition	2.75 x 21 inch	4.00 x 19 inch

Gold Star models

	Front	Rear
Touring	3.00 x 21 inch	3.25 x 19 inch, 350 cc (3.50 inch, 500 cc)
Scrambles	3.00 x 21 inch	4.00 x 19 inch
Trials	2.75 x 21 inch	4.00 x 19 inch
Racing	3.00 x 19 or 21 inch	3.25 x 19 inch (3.50 inch, 500 cc)

Tyre pressures (psi)

	Front	Rear
M20	17	22
M21	16	18
M33	17	18
B31	17	23
B32	20	16
B33	17	19
B34	22	16
B32GS (Touring)	21	22
B34GS (Touring)	21	18

Tyre pressures for use in competition events vary according to individual requirements and are therefore not given.

1 General description

As can be seen from the Specifications chart, all models are fitted with 19 inch rear wheels and with the exception of some Gold Star models, 19 inch front wheels. The tyre section of each wheel differs from model to model, depending on the weight of the machine and the use for which it was originally prepared.

Single leading shoe drum brakes are fitted to both wheels, of which the rear brake is 7 inches in diameter. The front brake is of 7 or 8 inches depending on the model and year. In addition a special 190 mm front brake is fitted to late 500 cc Gold Star models.

Throughout their entire production from 1954, all models, with the exception of M series machines, have quickly detachable rear wheels.

2 Front wheel: examination and renovation

1 Place the machine on the centre stand so that the front wheel is raised clear of the ground. Spin the wheel and check for rim alignment. Small irregularities can be corrected by tightening the

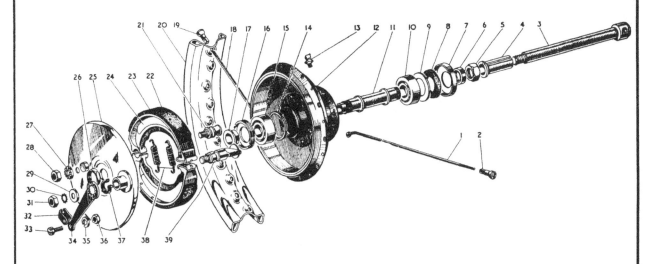

Fig. 7.1. Front hub and wheel - component parts

1	Spoke (20 off)	11	Spindle sleeve	21	Fulcrum
2	Nipple (20 off)	12	Front hub and drum	22	Brake lining rivet (14 off)
3	Wheel spindle	13	Grease nipple	23	Lining (2 off)
4	Distance piece	14	Thrust washer	24	Brake shoe (2 off)
5	Nut	15	Hub bearing	25	Backplate
6	Distance piece	16	Bearing retainer	26	Nut
7	Retaining cup	17	Spacer	27	Serrated washer
8	Felt washer	18	Spoke (20 off)	28	Nut
9	Retaining ring	19	Nipple (20 off)	29	Washer
10	Hub bearing	20	Wheel rim	30	Washer

31 Nut
32 Shackle
33 Screw
34 Brake actuating arm
35 Washer
36 Nut
37 Oil hole clip
38 Brake shoe (2 off)
39 Brake cam

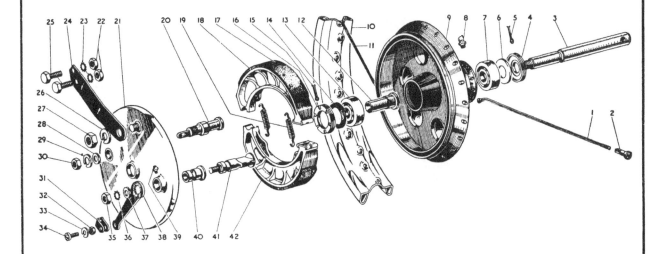

Fig. 7.2. Front hub and wheel - Gold Star - component parts

1 Spoke (20 off)
2 Nipple (40 off)
3 Wheel spindle
4 Bearing retainer
5 Split pin
6 Oil retainer
7 Wheel bearing
8 Grease nipple
9 Hub and drum
10 Wheel rim
11 Spoke (20 off)
12 Spindle sleeve
13 Wheel bearing
14 Oil retainer
15 Split pin
16 Retaining ring
17 Brake shoe rivet (14 off)
18 Brake shoe lining (2 off)
19 Brake shoe spring (2 off)
20 Fulcrum
21 Backplate
22 Nut (2 off)
23 Washer (2 off)
24 Torque arm
25 Bolt (2 off)
26 Spring washer
27 Nut
28 Washer
29 Spring washer
30 Nut
31 Shackle
32 Nut
33 Spring washer
34 Screw
35 Nut
36 Washer
37 Washer
38 Brake actuating arm
39 Grease nipple
40 Bush
41 Cam
42 Front brake shoe

3.1a Loosen the spindle nut to allow ...

3.1b ... removal of the brake back plate

3.3a Removal of operating cam after ...

spokes in the affected area, although a certain amount of
experience is necessary if over-correction is to be avoided. Any
'flats' in the wheel rim should be evident at the same time. These
are more difficult to remove with any success and in most cases
the wheel will have to be rebuilt on a new rim. Apart from the
effect on stability, especially at high speeds, there is much
greater risk of damage to the tyre beads and walls if the machine
is ridden with a deformed wheel.
2   Check for loose or broken spokes. Tapping the spokes is the
best guide to tension. A loose spoke will produce a quite
different note and should be tightened by turning the nipple in
an anti-clockwise direction. Always re-check for run-out by
spinning the wheel again.

3   Front brake assembly: examination, renovation and reassembly

1   The front wheel can be removed and the brake assembly
complete with brake plate detached by following the procedure
given in Chapter 6, Section 2, paragraphs 2 and 3.
2   Before dismantling the brake assembly, examine the
condition of the brake linings. If they are wearing thin or unevenly, they must be replaced.
3   To remove the brake shoes from the brake plate, position
the operating cam so that the shoes are in the fully-expanded
position and pull them apart whilst lifting them upwards, in the
form of a 'V'. When they are clear of the brake plate the return
springs should be removed and the shoes separated.
4   It is possible to replace the brake linings fitted with rivets and
not bonded on, as is the current practice. Much will depend on
the availabilty of the original type of linings; service-exchange
brake shoes with bonded-on linings may be the only practical
form of replacement.
5   If new linings are fitted by rivetting, it is important the
that the rivet heads are countersunk, otherwise they will rub on
the brake drum and be dangerous. Make sure the lining is in very
close contact with the brake shoe during the rivetting operation;
a small 'C' clamp of the type used by carpenters can often be used
to good effect until all the rivets are in position. Finish off by
chamfering off the end of each shoe, otherwise fierce brake grab
may occur due to the pick-up of the leading edge of each lining.
6   Before replacing the brake shoes, check that the brake
operating cam is working smoothly and not binding in its
pivot. The cam can be removed for greasing by unscrewing the
nut on the end of the brake operating arm and drawing the arm
off so that the cam and spindle can be withdrawn from the inside
of the brake plate.

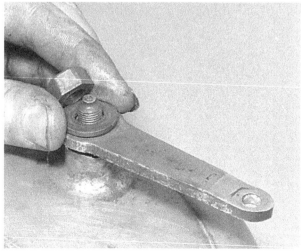

3.3b ... detaching the brake arm may assist ...

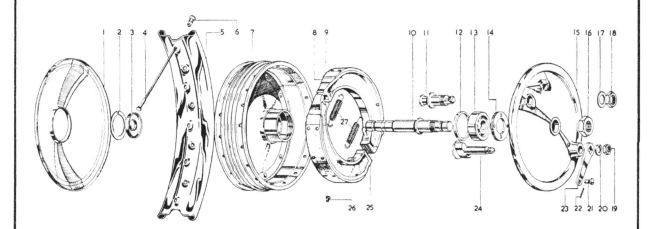

Fig. 7.3. Front hub and wheel 'B' models - component parts

1 Cover plate
2 Circlip
3 Dust cover
4 Spoke (40 off)
5 Wheel rim
6 Nippie (40 off)
7 Hub and drum

8 Brake shoe (2 off)
9 Brake lining (2 off)
10 Wheel spindle
11 Fulcrum
12 Spacer
13 Wheel bearing (2 off)
14 Lockring

15 Backplate
16 Nut
17 Washer
18 Nut
19 Nut
20 Washer

21 Clevis pin
22 Split pin
23 Brake actuating lever
24 Brake cam
25 Rivet (16 off)
26 Spring (2 off)

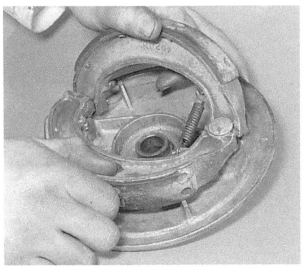

3.3c ... pulling the brake shoes from position

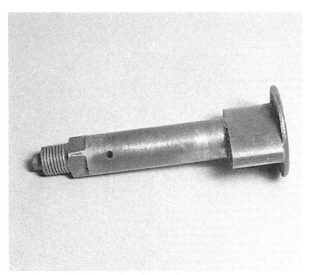

3.6 Check condition of brake cam and spindle

4.1a Drum side wheel bearing is held by a retainer ring

4.1b Some rings are marked to identify thread

4.1c Some bearings are retained by a circlip

7   Check the inner surface of the brake drum, on which the brake shoes bear. The surface should be smooth and free from indentations, or reduced braking efficiency is inevitable. Remove all traces of brake lining dust and wipe the surface with a rag soaked in petrol to remove any traces of grease or oil.

8   To reassemble the brake shoes on the brake plate, fit the return springs and force the shoes apart, holding them in a 'V' formation. If they are now located with the brake operating arm and pivot, they can usually be snapped into position by pressing downward. Do not use excessive force, or the shoes may be distorted permanently.

4   Wheel bearings: examination and replacement

1   When the brake plate complete with brake assembly has been removed, the bearing retainer within the brake drum will be exposed. Where fitted, remove the securing pin before attempting to loosen the ring either with a peg spanner or a soft brass punch. The retainer ring has either a left-hand or a right-hand thread, depending on the hub type. Most hubs have retaining rings using a left-hand thread. This does not apply to 8 inch half-width steel hubs. If there is any doubt experimentation will have to be adopted. Left-hand thread rings are often stamped LH. The ball journal on the right-hand side is extracted by striking the left-hand end of the front wheel spindle with a rawhide mallet, or if the spindle is not of the captive type, by using a suitable diameter drift to drive through the centre of the left-hand bearing. The right-hand bearing will be displaced, together with the hollow sleeve through which the wheel spindle passes, if the latter is of the detachable type. The left-hand bearing can then be driven out from the right-hand side of the hub. Here again, some hubs are fitted with retainer rings on the left-hand side of the hub which may be secured by a split pin. In some cases a circlip is used to retain the bearing.

2   Remove all the old grease from the hub and bearings, giving the latter a final wash in petrol. Check the bearings for play or signs of roughness when they are turned. If there is any doubt about their condition, play safe and renew them. A new bearing has no discernible play.

3   Before replacing the bearings, first pack the hub with new, high melting point grease. Then grease both bearings and drive them back into position, not forgetting any distance piece, hollow sleeve or shim washers that were fitted originally. Make sure the bearing retainer is tight and that the dust cover is located correctly. The bearing retainer performs the dual role of preventing grease from entering the brake drum, thereby reducing braking efficiency.

4.1d Use spindle to drift out oil seal (where fitted) and ...

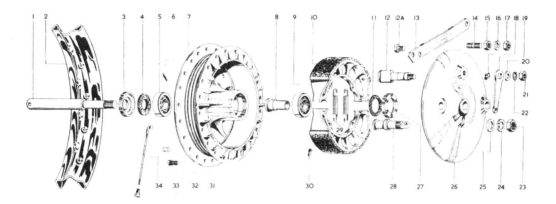

Fig. 7.4. Front full-width hub Gold Star models only - component parts

1 Front wheel spindle	10 Lining (2 off)
2 Wheel rim	11 Oil retaining washer
3 Lockring	12 Brake cam
4 Oil retaining washer	12a Bolt (2 off)
5 Wheel bearing (2 off)	13 Torque arm
6 Split pin	14 Stud
7 Hub and drum	15 Nut
8 Spindle sleeve	16 Spring washer
9 Brake shoe (2 off)	17 Nut

18 Grease nipple	27 Lockring
19 Brake actuating arm	28 Fulcrum
20 Washer	29 Spring (2 off)
21 Nut	30 Rivet (20 off)
22 Washer	31 Dowel (5 off)
23 Nut	32 Screw (5 off)
24 Spring washer	33 Nipple (40 off)
25 Washer	34 Spoke (40 off)
26 Backplate	

4.1e ... the bearing from the hub

4.3 Socket may be used to drift new bearings home

5.1 Lug on fork must engage with brake back plate

6.2a Unscrew the four sprocket nuts and ...

6.2b ... detach the torque arm at brake plate

4   There is no means of adjusting wheel bearings of the ball journal type. If play is evident, the bearings have reached the end of their useful service life.

5   Front wheel: reassembly and replacement

1   Replace the front brake assembly in the brake drum and align the front wheel so that the projection on the brake plate engages with the peg on the right-hand fork leg. The importance of this cannot be overstressed because the anchorage of the front brake is dependent on the correct location of these parts.
2   Some models have an alternative torque arm arrangement. In this case the torque arm must be slipped over the end of the stud in the brake plate before the brake spindle is inserted and tightened.
3   Replace the front wheel spindle, if of the detachable type, or replace the split clamps at the extreme end of each fork leg. Before tightening the pinch bolt that secures the front wheel spindle, depress the forks several times so that the left-hand fork leg can position itself correctly on the distance bush. (detachable models only).
4   Reconnect the front brake cable and check that the brake functions correctly. If the connection through the brake operating arm is made by means of a clevis pin, a new split pin must be used to secure the clevis pin in position. Re-check the wheel spindle nut, pinch bolt or the fork end bolts for tightness, also the nuts that secure the separate torque arm (if fitted).

6   Rear wheel: examination, removal and renovation

1   Before removing the rear wheel, check for rim alignment, damage to the rim and loose or broken spokes by following the procedure that applies to the front wheel examination, in the preceding Section. The method adopted when removing the rear wheel depends upon whether or not it is quickly detachable and if quickly detachable, which type of hub is employed. With quickly detachable wheels, disconnection of the rear chain and removal of the sprocket is not required. Furthermore, on machines with the brake drum on the sprocket side, the drum does not require removal either.

Alloy full width hub 1955 - 60 B31 - 33
2   To remove the rear wheel without disconnecting the chain, place the machine on the centre stand so that the rear wheel is

6.3a Unscrew spindle to release wheel

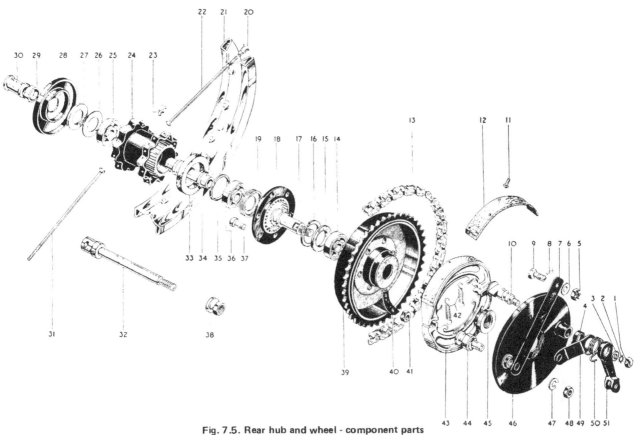

Fig. 7.5. Rear hub and wheel - component parts

1	Nut	14	Hub bearing	27	Felt washer	40
2	Washer	15	Shim	28	End cover	41
3	Washer	16	Spring ring	29	Spacer	42
4	Oil hole clip	17	Bearing sleeve	30	Spacer	43
5	Nut	18	Drive flange	31	Spoke (20 off)	44
6	Washer	19	Lockring	32	Wheel spindle	45
7	Torque arm	20	Nipple (40 off)	33	Rubber washer	46
8	Split pin	21	Rim	34	Sleeve	47
9	Bolt	22	Spoke (20 off)	35	Thrust washer	48
10	Cam	23	Grease nipple	36	Wheel bearing	49
11	Rivet (14 off)	24	Rear hub	37	Bolt (6 off)	50
12	Lining (2 off)	25	Hub bearing	38	Nut	51
13	Final drive chain	26	Retainer	39	Drum and sprocket	

40 Double tab washer (3 off)
41 Nut (6 off)
42 Spring (2 off)
43 Brake shoe (2 off)
44 Fulcrum
45 Collar
46 Backplate
47 Spring washer
48 Nut
49 Outrigger arm
50 Return spring
51 Brake actuating arm

raised clear of the ground. Take off the rear brake cable by disconnecting it from the brake operating arm and the brake plate and remove the nut and washer that retain the rear brake torque arm to the brake plate.

3   Unscrew and remove the four nuts that secure the rear wheel sprocket to the hub. Models fitted with a fully-enclosed final drive chain have a rubber plug in the outer face of the chaincase which, when removed, will give access to the nuts (one by one) as the wheel is rotated. A box or socket spanner will be necessary to remove the nuts.

4   Apply a spanner or tommy bar (early models) to the end of the rear wheel spindle until it can be withdrawn completely. Remove the distance piece fitted between the hub and the right-hand side of the machine and pull the wheel to the right so that the studs will clear the sprocket boss. The wheel can now be drawn downwards and to the rear until it clears the frame and rear mudguard.

5   Do not disturb the wheel nut on the left-hand side of the machine. This retains the sprocket and hollow spindle into which the wheel spindle is threaded. It need be removed only if the wheel complete with sprocket attached has to be withdrawn from the frame, or the sprocket itself after the wheel is removed.

6.3b Pull wheel towards right-hand side to free

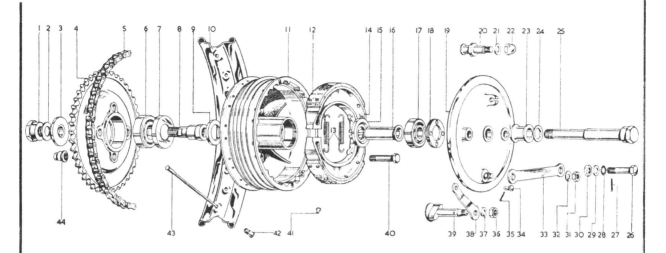

Fig. 7.6. Full width rear hub and wheel - component parts

1 Nut	12 Brake lining (2 off)	23 Distance piece	34 Clevis pin
2 Washer	13 Brake spring (2 off)	24 Washer	35 Split pin
3 Distance collar	14 Brake shoe (2 off)	25 Wheel spindle	36 Nut
4 Rear sprocket	15 Bearing collar	26 Bolt	37 Spring washer
5 Final drive chain	16 Distance piece	27 Split pin	38 Brake actuating lever
6 Wheel bearing	17 Wheel bearing	28 Washer	39 Brake cam
7 Grease retainer	18 Lockring	29 Washer	40 Bolt (4 off)
8 Fixed spindle	19 Backplate	30 Nut	41 Rivet (16 off)
9 Dust cover	20 Fulcrum	31 Nut	42 Nipple (40 off)
10 Wheel rim	21 Washer	32 Spring washer	43 Spoke (40 off)
11 Rear hub and drum	22 Domed nut	33 Torque arm	44 Sprocket nut (4 off)

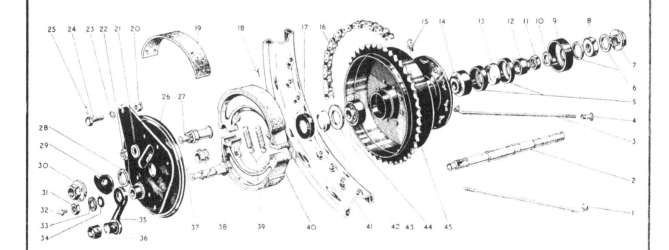

Fig. 7.7. Rear hub and wheel rigid models - component parts

1 Spoke (20 off)	13 Felt washer	24 Bolt	35 Brake actuating lever
2 Wheel spindle	14 Bearing	25 Split pin	36 Bush
3 Spoke (20 off)	15 Grease nipple	26 Backplate	37 Locknut
4 Nipple (40 off)	16 Rear drive chain	27 Fulcrum	38 Cam
5 Retaining cup (2 off)	17 Retaining cup	28 Washer	39 Spring (2 off)
6 Washer (2 off)	18 Rivets (14 off)	29 Snail cam	40 Brake shoe (2 off)
7 Nut	19 Lining (2 off)	30 Spindle nut	41 Wheel rim
8 Locknut	20 Nut	31 Nut	42 Felt washer
9 Cap	21 Washer	32 Grease nipple	43 Retaining washer
10 Washer	22 Nut	33 Washer	44 Wheel bearing
11 Nut	23 Washer	34 Washer	45 Hub and drum
12 Adjusting sleeve			

Non-quickly detachable hub, rigid frame, M Series

6   Place the machine on the rear stand so that the rear wheel is clear of the ground. Disconnect the rear chain by removal of the spring line and run the chain off the rear wheel sprocket. Unscrew the brake torque arm nut, after removing the split pin.

7   Disconnect the rear brake rod by unscrewing the knurled adjuster nut and then applying the brake so that the rod leaves the trunnion in the brake operating arm.

8   Slacken the two nuts on the rear mudguard and hinge the guard upwards. Loosen the two wheel spindle nuts and slide the wheel out from the open fork ends.

Quickly detachable half-width steel hubs. Plunger frame and all Gold Star models

9   Slacken the nuts at the ends of the slotted mudguard stay ends and hinge up the mudguard. (plunger models only).

10 Unscrew the wheel spindle from the right-hand side of the machine, using a tommy bar. On plunger models the small locknut on the extreme left-hand end of the spindle will first require removal. Draw out the wheel spindle and remove the wheel spacer. The wheel can now be moved over to the right-hand side of the machine until the splined drive boss leaves the splines in the brake drum hub.

7 Brake drum: removal

Quickly-detachable steel hub wheel only

1   Disconnect the rear chain by removal of the spring link, and run the chain off the sprocket.

2   Disconnect the brake rod by removing the knurled adjuster nut and applying the brake so that the rod end leaves the trunnion in the brake arm.

3   Remove the torque arm from the brake back plate by unscrewing the retaining nut.

4   Loosen and remove the large spindle nut, and at the same time support the weight of the hub, drum and sprocket.

8 Rear brake assembly: examination, renovation and re-assembly

1   The rear brake assembly can be withdrawn from the brake drum after the rear wheel has been removed, or in the case of quickly-detachable steel hub wheels, after the drum has been detached from the frame.

2   Follow an identical procedure for dismantling and reassembly to that relating to the front brake, as described in Section 3 of this Chapter. The brake assemblies are identical, even to the extent of interchangeable brake plates on some of the earlier models.

9 Rear wheel bearings: examination and replacement

1   The rear wheel bearings are of the non-adjustable ball journal type. They are a drive fit within the hub and can be removed by following the procedure adopted for the removal of the front wheel bearings, as described in Section 4. Here again the bearing retainer ring (where fitted) has a right or left-hand thread depending on the model.

2   The greasing and reassembly procedure is identical to that described in Section 4, paragraphs 3 and 4.

3   On quickly detachable steel hub wheels a third bearing is fitted, which is located in the brake drum carrying hub. This bearing can be driven out from the drum side in the usual manner.

10 Front and rear brakes: adjustment

Full width hub with screw-type fulcrum pin

1   The brakes fitted to 1956 B31 and B33 models incorporate

8.1 Sprocket spindle is held by one nut

a cone-type brake fulcrum pin, which allows the static position of the brake shoes to be altered in relation to the brake drum. Before the normal brake adjustment is carried out by means of the cable or rod adjusters, fulcrum adjustment should be made.

2   Place the machine on the centre stand so that the wheel in question is clear of the ground. Place a spanner on the square end of the pin where it protrudes from the brake back plate and screw it inwards slowly. Rotate the wheel whilst turning the fulcrum pin. When the brake begins to bind, slacken the pin just sufficiently to allow the wheel to rotate freely.

3   If required, the front brake cable may be adjusted, or the rear brake rod adjusted, by means of the knurled nut, to make the final adjustment.

7 inch steel hub front brake

4   Some steel hub front brakes are fitted with an adjustable fulcrum pin that registers in an elongated hole in the brake plate. This allows adjustment of the brake shoes so that they are positioned exactly concentrically with the brake drum. If the shoes have been removed or renewed, or braking efficiency is suspect, adjustment of the fulcrum pin should be made.

5   Slacken off the nut on the fulcrum pin and apply the front brake fully. The shoes will centre themselves automatically as they are expanded against the drum. Without releasing the brake, tighten the fulcrum nut.

6   Adjustment of the brake should be carried out as described below.

All models

7   The front brake adjuster is located on the front brake plate or on a lug welded to the fork leg. Some models have an additional adjuster built-in to the end of the brake operating lever on the handlebars.

8   The front brake should be adjusted so that the wheel is free to revolve before pressure is applied to the handlebar lever. It should be applied fully before the handlebar lever touches the handlebars. Make sure the adjuster locknut is tight after the correct adjustment has been made.

9   The rear brake is adjusted by means of an adjusting nut on the end of the brake operating rod, or when a cable is used, by means of an adjuster attached to the brake plate of a similar type to that employed for the front brake. Adjustment is largely a matter of personal choice, but excessive pedal travel should be avoided before the brake is applied fully.

10 Efficient brakes depend on good leverage of the brake operating arm. The angle between the brake operating arm and the cable or rod should never exceed 90° when the brake is applied fully.

11 Rear wheel sprocket: removal, examination and replacement

1 On all models excluding the machines fitted with full-width alloy hubs, the rear wheel sprocket is an integral part of the rear brake drum. It follows that if the sprocket requires renewal the brake drum must be renewed and vice-versa. On machines using a full-width alloy hub the rear wheel sprocket is attached to the hub by four shouldered nuts that engage with studs threaded into the hub. The centre of the sprocket carries a short, hollow spindle into which the rear wheel spindle is threaded; the wheel must be removed from the frame and the chain disconnected, before the sprocket can be separated, unless the wheel is first removed separately as described in Section 6.
2 Models fitted with a fully enclosed final drive chain must have both halves of the chaincase removed, before the chain can be detached from the sprocket.
3 Check the condition of the sprocket teeth. If they are hooked, chipped or badly worn, the sprocket must be renewed.
4 It is bad practice to renew one sprocket on its own. The final drive sprockets should always be replaced as a pair and a new chain fitted, otherwise rapid wear will necessitate even earlier replacement next time.

12 Final drive chain: examination, lubrication and adjustment

1 Except on a few models, the final drive chain does not have the benefit of full enclosure or positive lubrication that is afforded to the primary drive chain. In consequence, it will require attention from time to time, particularly when the machine is used on wet or dirty roads.
2 Chain adjustment is correct when there is approximately ¾ inch play in the middle of the run. Always check at the tightest spot on the chain run, under load.
3 If the chain is too slack, adjustment is effected by slackening the rear wheel spindle and the torque arm (when fitted), then drawing the wheel backwards by means of the chain adjusters at the end of the rear fork. Make sure each adjuster is turned an equal amount, so that the rear wheel is kept centrally-disposed within the frame. When the correct adjusting point has been reached, push the wheel hard forward to take up any slack, then tighten the spindle, not forgetting the torque arm nut, if fitted. Re-check the chain tension and the wheel alignment, before the final tightening of the spindle and nuts.
4 Application of engine oil from time to time will serve as a satisfactory form of lubrication, but it is advisable to remove the chain every 2000 miles (unless it is enclosed within a chaincase, in which case every 5000 miles should suffice) and clean it in a bath of paraffin before immersing it in a special chain lubricant such as "Linklyfe" or "Chainguard". These latter types of lubricant achieve better and more lasting penetration of the chain links and rollers and are less likely to be thrown off when the chain is in motion.
5 To check whether the chain is due for replacement, lay it lengthwise in a straight line and compress it, so that all play is taken up. Anchor one end and then pull on the other, to stretch the chain in the opposite direction. If the chain extends by more than ¼ inch per foot, replacement is necessary.
6 When replacing the chain, make sure the spring link is positioned correctly, with the closed end facing the direction of travel. Reconnection is made easier if the ends of the chain are pressed into the rear wheel sprocket.

13 Rear wheel: replacement

1 The rear wheel is replaced in the frame by reversing the dismantling procedure described in Section 6.
2 The drive splines on quickly detachable steel hub wheels should be lubricated with heavy grease if they are dry. Ensure that any distance pieces fitted between the hub or brake back-

plate are replaced. Check that the brake torque arm has been replaced and that all bolts and nuts are tightened fully.
3 On full-width alloy hub models, if the rear wheel sprocket has been removed either with or without the rear wheel attached, make sure that the four shouldered nuts retaining the sprocket to the hub are tightened fully. If these nuts work loose, they will place a shear stress on the retaining studs, leading to their early failure.

14 Wheel balance

1 On any high performance machine it is important that the front wheel is balanced, to offset the weight of the tyre valve. If this precaution is not observed, the out-of-balance wheel will produce an unpleasant hammering that is felt through the handlebars at speeds from approximately 50 mph upwards.
2 To balance the front wheel, place the machine on the centre stand so that the front wheel is well clear of the ground and check that it will revolve quite freely, without the brake shoes rubbing. In the unbalanced state, it will be found that the wheel always comes to rest in the same position, with the tyre valve in the six o'clock position. Add balance weights to the spokes diametrically opposite the tyre valve until the the tyre valve is counterbalanced, then spin the wheel to check that it will come to rest in a random position on each occasion. Add or subtract weight until perfect balance is achieved.
3 Only the front wheel requires attention. There is little point in balancing the rear wheel (unless both wheels are completely interchangeable) because it will have little noticeable effect on the handling of the machine.
4 Balance weights of various sizes that will fit around the spoke nipples were originally available from BSA Motor Cycles. If difficulty is experienced in obtaining them, lead wire or even strip solder can be used as an alternative, kept in place with insulating tape.

15 Tyres: removal and replacement

1 At some time or other the need will arise to remove and replace the tyres, either as the result of a puncture or because replacements are necessary to offset wear. To the inexperienced, tyre changing represents a formidable task yet if a few simple rules are observed and the technique learned, the whole operation is surprisingly simple.
2 To remove the tyre from either wheel, first detach the wheel from the machine by following the procedure in Chapters 5.2 or 6.6, depending on whether the front of the rear wheel is involved. Deflate the tyre by removing the valve insert and when it is fully deflated, push the bead from the tyre away from the wheel rim on both sides so that the bead enters the centre well of the rim. Remove the locking cap and push the tyre valve into the tyre itself.
3 Insert a tyre lever close to the valve and lever the edge of the tyre over the outside of the wheel rim. Very little force should be necessary; if resistance is encountered it is probably due to the fact that the tyre beads have not entered the well of the wheel rim all the way round the tyre.
4 Once the tyre has been edged over the wheel rim, it is easy to work around the wheel rim so that the tyre is completely free on one side. At this stage, the inner tube can be removed.
5 Working from the other side of the wheel, ease the other edge of the tyre over the outside of the wheel rim that is furthest away. Continue to work around the rim until the tyre is free completely from the rim.
6 If a puncture has necessitated the removal of the tyre, re-inflate the inner tube and immerse it in a bowl of water to trace the source of the leak. Mark its position and deflate the tube. Dry the tube and clean the area around the puncture with a petrol soaked rag. When the surface has dried, apply rubber solution and allow this to dry before removing the backing from the patch and applying the patch to the surface.

Tyre changing sequence - tubed tyres

 A Deflate tyre. After pushing tyre beads away from rim flanges push tyre bead into well of rim at point opposite valve. Insert tyre lever adjacent to valve and work bead over edge of rim.

 B Use two levers to work bead over edge of rim. Note use of rim protectors

 C Remove inner tube from tyre

 D When first bead is clear, remove tyre as shown

E When fitting, partially inflate inner tube and insert in tyre

F Work first bead over rim and feed valve through hole in rim. Partially screw on retaining nut to hold valve in place.

 G Check that inner tube is positioned correctly and work second bead over rim using tyre levers. Start at a point opposite valve.

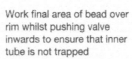

H Work final area of bead over rim whilst pushing valve inwards to ensure that inner tube is not trapped

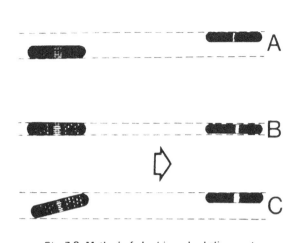

Fig. 7.8. Method of checking wheel alignment

A & C   Incorrect   B Correct

7   It is best to use a patch of the self-vulcanising type, which
will form a very permanent repair. Note that it may be necessary
to remove a protective covering from the top surface of the patch,
after it has sealed in position. Inner tubes made from synthetic
rubber may require a special type of patch and adhesive, if a
satisfactory bond is to be achieved.

8   Before replacing the tyre, check the inside to make sure the
agent that caused the puncture is not trapped. Check the
outside of the tyre, particularly the tread area, to make sure
nothing is trapped that may cause a further puncture.

9   If the inner tube has been patched on a number of past
occasions, or if there is a tear or large hole, it is preferable to
discard it and fit a replacement. Sudden deflation may cause an
accident, particularly if it occurs with the front wheel.

10 To replace the tyre, inflate the inner tube sufficiently for it
to assume a circular shape but only just. Then push it into the
tyre so that it is enclosed completely. Lay the tyre on the wheel
at an angle and insert the valve through the rim tape and the hole
in the wheel rim. Attach the locking cap on the first few threads,
sufficient to hold the valve captive in its correct location.

11 Starting at the point furthest from the valve, push the tyre
bead over the edge of the wheel rim until it is located in the
central well. Continue to work around the tyre in this fashion
until the whole of one side of the tyre is on the rim. It may be
necessary to use a tyre lever during the final stages.

12 Make sure there is no pull on the tyre valve and again
commencing with the area furthest from the valve, ease the other
bead of the tyre over the edge of the rim. Finish with the area

close to the valve, pushing the valve up into the tyre until the
locking cap touches the rim. This will ensure the inner tube is
not trapped when the last section of the bead is edged over the
rim with a tyre lever.

13 Check that the inner tube is not trapped at any point.
Re-inflate the inner tube, and check that the tyre is seating
correctly around the wheel rim. There should be a thin rib
moulded around the wall of the tyre on both sides, which should
be equidistant from the wheel rim at all points. If the tyre is
unevenly located on the rim, try bouncing the wheel when the
tyre is at the recommended pressure. It is probable that one of
the beads has not pulled clear of the centre well.

14 Always run the tyres at the recommended pressures and
never under or over-inflate. The correct pressures are given in the
Specifications Section . If a pillion passenger is carried,
increase the rear tyre pressure only to 28 psi.

15 Tyre replacement is aided by dusting the side walls, particularly in the vicinity of the beads, with a liberal coating of
french chalk. Washing-up liquid can also be used to good effect,
but this has the disadvantage of causing the inner surfaces of the
wheel rim to rust.

16 Never replace the inner tube and tyre without the rim tape
in position. If this precaution is overlooked there is good chance
of the ends of the spoke nipples chafing the inner tube and
causing a crop of punctures.

17 Never fit a tyre that has damaged tread or side walls. Apart
from the legal aspects, there is a very great risk of blow-out,
which can have serious consequences on any two-wheel vehicle.

18 Tyre valves rarely give trouble, but it is always advisable to
check whether the valve itself is leaking before removing the
tyre. Do not forget to fit the dust cap, which forms an effective
second seal.

16 Tyre valve dust caps

1   Tyre valve dust caps are often left off when a tyre has been
replaced, despite the fact that they serve an important two-fold
function. Firstly they prevent dirt or other foreign matter from
entering the valve and causing the valve to stick open when the
tyre pump is next applied. Secondly, they form an effective
second seal so that in the event of the tyre valve sticking, air will
not be lost.

2   Isolated cases of sudden deflation at high speeds have been
traced to the omission of the dust cap. Centrifugal force has
tended to lift the tyre valve off its seating and because the
dust cap is missing, there has been no second seal. Racing inner
tubes contain provision for this happening because the valve
inserts are fitted with stronger springs, but standard inner tubes
do not, hence the need for the dust cap.

3   Note that when a dust cap is fitted for the first time, the
wheel may have to be rebalanced.

17 Fault diagnosis: wheels, brakes and tyres

Symptom	Cause	Remedy
Handlebars oscillate at low speeds	Buckle or flat in wheel rim	Check rim alignment by spinning. Correct by retensioning spokes or by having wheel rebuilt on new rim.
	Tyre not straight on rim	Check tyre alignment.
Machine lacks power and accelerates poorly	Brakes binding	Warm brake drums provide best evidence. Re-adjust brakes.
Brakes grab, even when applied gently	Ends of brake shoes not chamfered	Chamfer with file.
	Elliptical brake drum	Lightly skim in lathe (specialist attention required).
Brake pull-off sluggish	Brake cam binding in housing	Free and grease.
	Weak brake shoe springs	Renew, if springs not displaced

Harsh transmission	Worn or badly adjusted chains	Renew or adjust, as necessary.
	Hooked or badly worn sprockets	Renew as a pair.
	Rear wheel sprocket nuts loose	Check and tighten.
Middle of tyre treads wear rapidly	Tyres over-inflated	Check and readjust pressures.
Edges of tyre treads wear rapidly	Tyres under inflated	Check and increase pressures.
Forks hammer at high speeds	Front wheel not balanced	Balance wheel by adding balance weights.

Chapter 8 Electrical system

Contents

General description	1	only)	9
Dynamo: checking the output	2	Battery: examination and maintenance	10
Dynamo: removal	3	Battery: charging procedure	11
Dynamo drive gear: removal and inspection	4	Headlamp: replacing bulbs and adjusting beam height ...	12
Dynamo: lubricating the bearings	5	Tail and stop lamps: replacing bulbs	13
Voltage regulator: function and adjustments	6	Wiring: layout and examination	14
Voltage regulator: checking the settings	7	Headlamp switch	15
Alternator: checking the output	8	Ignition switch (alternator models)	16
Selenium rectifier: general description (alternator models		Fault diagnosis: electrical system	17

Specifications

Battery
Make	Lucas
Type	Lead acid PU7E
Capacity	12 amp/hr
Voltage	6

Dynamo
Make	Lucas
Type	E3L
Output	60 watts
Voltage	6

Regulator
Make	Lucas
Type	MCR 2

Models B31-34 post 1957

Alternator
Make	Lucas
Type	Six coil, permanent rotating magnet
Output	6 volts, 60 watts

Rectifier
Make	Lucas
Type	Full wave, selenium

All models have a positive (+ uc) earth system.

Bulbs
	Main	Pilot	Tail/stop	Instrument
All models	30/24w	3w	6/18w	2.2w

All bulbs rated at 6 volts.

1 General description

All the models covered in this manual, excluding the 1958-60 B31 and B33 models, utilise a direct current lighting system powered by a battery and dynamo. The dynamo is incorporated in a Magdyno unit in which the magneto body serves as a cradle for the independent dc dynamo generator which is driven by gears via the magneto armature shaft.

The electrical system for all the models is fundamentally the same, differing from one to another only in minor detail.

The direct current lighting system on B31 and B33 machines was eventually superseded by an ac system, powered by an alternator mounted on the extreme left-hand end of the crankshaft. The current supplied by the alternator is rectified to dc in order to supply the battery and the ignition and lighting systems.

All machines utilise a positive (+) earth electrical system.

2 Dynamo: checking the output

1 The output from the dynamo can be checked by removing the electrical lead from the end cover and bridging both terminals together, to form one point for the connection of a dc voltmeter. The other connection from the meter (positive) should be attached to the dynamo casing or some other convenient earthing point of the machine. Running on open circuit, the dynamo should give a voltmeter reading of 7 - 8 volts, when the engine is started and is running at a fast tickover.

2 If a voltmeter is not available, visual evidence of a charge can be obtained by connecting a headlamp bulb in the same manner as the voltmeter. Use a 12 volt bulb, otherwise the 6 volt bulb will blow, due to the excess voltage developed when the dynamo is running on open circuit. The bulb should glow quite brightly when the engine is started.

3 If there is no evidence of a charge, check that the dynamo armature is rotating. A loose driving pinion has been known to give the illusion of a faulty dynamo.
dynamo.

4 A dynamo is now regarded as an obsolete instrument and there is no longer the possibility of obtaining a service exchange replacement from the manufacturer at moderate cost. Some of the larger dealers still offer a form of service exchange replacement; the alternative is to entrust the repair to a dynamo repair specialist, who will have the facilities for rewinding the armature or the field coil - the two parts of a dynamo most likely to break down.

3 Dynamo: removal

1 The dynamo can be removed from position on the Magdyno after disconnecting the two leads from the end cover and removal of the dynamo clamping strap bolt. In addition, the small bevelled nut must be removed from the stud that projects through the front of the dynamo drive cover plate.

4 Dynamo drive gear: removal and inspection

1 The dynamo is driven by means of a large fibre gear mounted concentrically with the magneto armature shaft. Drive

3.1 Dynamo is retained by strap and locating stud

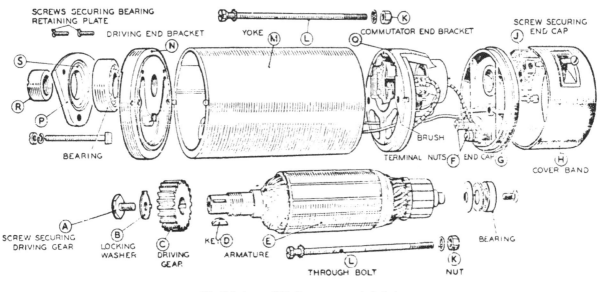

Fig. 8.1. Lucas E3L Dynamo - exploded view

4.4 Drive cover is secured by four countersunk screws

is transmitted from the magneto shaft by a spring star plate, the centre of which is retained on the shaft and the 'fingers' of which press against the recessed face of the fibre gear. The fibre gear intermeshes with a small diameter steel pinion on the dynamo armature shaft.

2   The spring plate is incorporated to act as a clutch, which will slip when a sudden heavy load is applied to the dynamo.

3   The dynamo drive rarely gives trouble unless the spring plate loses tension. The fibre gear will eventually wear, necessitating its replacement.

4   To gain access to the drive gear, the Magdyno must first be removed and the drive cover plate detached. The cover is retained by four screws. To remove the fibre drive gear, bend down the ear of the tab washer and loosen the centre nut. Remove the nut followed by the tab washer, spring plate, fibre gear pressure ring and the fibre gear.

5   When replacing the drive gear, the tension of the spring plate must be adjusted so that 4 - 10 lb ft torque must be applied to the magneto armature before the clutch slips. Adjustment of the spring plate may be made by turning the centre retaining bolt. The point at which the clutch slips can be ascertained by attaching a 12 inch arm to the magneto shaft, the end of which is connected with a spring balance. By holding the fibre wheel stationary the force required to induce slippage may be read off the spring balance (see accompanying diagram).

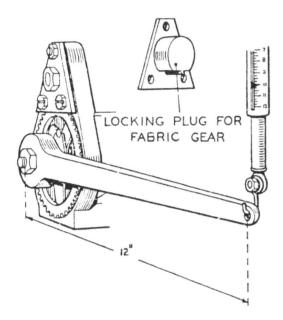

Fig. 8.2. Method of checking generator drive

5   Dynamo: lubricating the bearings

The dynamo is fitted with ball journal bearings at both ends of the armature. The ball journal bearings are packed with grease that will last until the machine is dismantled for a general overhaul. It is then opportune to repack the bearings with similar high melting point grease, provided no play is evident that will necessitate replacement of the bearings.

6   Voltage regulator: function and adjustments

1   The voltage regulator, which takes the form of a small, oblong box clamped to some convenient part of the machine (or housed in the toolbox), performs a dual function. It contains the cut-out, an electro-magnetic device that determines the point at which the dynamo is connected to the charging circuit and the point at which it is disconnected. If this provision was not included in the electrical system, the battery would discharge

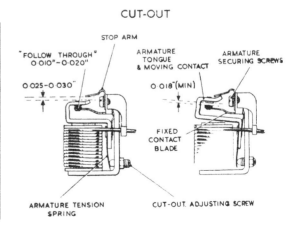

Fig. 8.3. Regulator and cut-out adjustment and setting

through the dynamo when it was stationary or running at low speeds, causing damage to both the battery and the field coil of the dynamo. As its name implies, the regulator also controls the output from the dynamo, when it is connected to the charging circuit, by means of another electromagnetic method. This explains the presence of the two separate coils in the regulator unit, each with its own set of contacts. The regulator matches the charge from the dynamo to the requirements of the battery, hence if the battery is discharged, a full charge will be given. If, on the other hand, the battery is fully charged, the charge rate will be cut down so that only a trickle charge is given until a heavy load is again placed on the battery.

2   The regulator is correctly adjusted during manufacture and further adjustments should not be required until the machine has seen a considerable amount of service. The parts most likely to require attention are the contacts, which should be cleaned with either fine emery cloth or an oilstone. Complete the operation by cleaning with methylated spirits, to remove all traces of dust and foreign matter.

3   It will probably be necessary to remove the armature plate to gain access to the contact points of the regulator coil, in which case the air gap between the core of the coil and the armature plate will have to be reset. This is accomplished by slackening off the locknut on the voltage adjusting screw and unscrewing the screw until it is clear of the flat spring that tensions the armature plate. Slacken the two screws that secure the armature plate to the main body of the regulator unit and insert a 0.015 in feeler gauge between the armature plate and the core of the regulator coil. Press the armature plate until it is squarely in contact with the feeler gauge and tighten the two retaining screws. Before the feeler gauge is withdrawn turn the voltage adjusting screw until it just touches the armature plate tension spring and tighten the locknut. Check, and if necessary reset the voltage adjusting screw, as described in Section 7 of this Chapter.

4   If the cutout points are dismantled for cleaning, they should be reset so that there is a gap of 0.025 in - 0.030 in between the stop arm and the moving contact, when the contact points are closed. In the open position there should be a minimum gap of 0.018 in between the points.

7   Voltage regulator: checking the settings

1   To check the electrical settings it is necessary to have a good quality voltmeter of the moving coil type, with a range of 0 - 20 volts dc. Connect the negative lead of the voltmeter to the D terminal of the regulator unit and the positive lead to the E terminal. Detach the negative lead from the battery (positive lead, early machines with negative earth). Start the engine and slowly increase the engine speed until the voltmeter needle 'kicks' then steadies. Note the reading and stop the engine.

2   The setting is correct if the voltage on open circuit is within the following limits:

Air temperature		Acceptable voltage range
10°C	50°F	7.7 − 8.1
20°C	68°F	7.6 − 8.0
30°C	86°F	7.5 − 7.9
40°C	104°F	7.4 − 7.8

If the voltage is outside the acceptable range, the regulating screw must be readjusted. Turn clockwise to raise the setting, or anti-clockwise to decrease it, noting that only a fraction of a turn may be necessary to achieve a marked change in the readings. When the setting is correct, tighten the locknut.

3   Adjustment should be effected within 30 seconds, otherwise the shunt winding will heat up and give rise to false settings. If the regulator unit is removed during these checking operations, make sure it is held in a similar position to that adopted on the machine.

4   Do not run the engine at more than half engine speed, otherwise the dynamo will build up a high voltage because it is running on open circuit.

6.1 Voltage regulator is often housed in toolbox

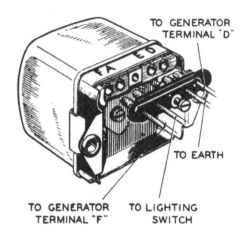

TO GENERATOR TERMINAL "D"

TO EARTH

TO GENERATOR TERMINAL "F"

TO LIGHTING SWITCH

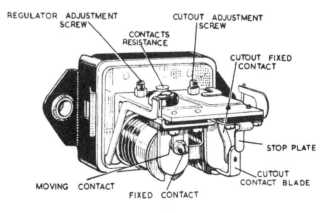

REGULATOR ADJUSTMENT SCREW

CUTOUT ADJUSTMENT SCREW

CONTACTS RESISTANCE

CUTOUT FIXED CONTACT

STOP PLATE

CUTOUT CONTACT BLADE

MOVING CONTACT

FIXED CONTACT

Fig. 8.4. Control box connections and internal layout

5   To check the electrical setting of the cut-out, connect the voltmeter to the D and E terminals,of the regulator, as previously, but do not detach the negative lead from the battery. Start the engine and increase the engine speed slowly until the cutout contacts close. Note the reading and stop the engine.

6   If the reading is outside the limits of 6.3 - 6.7 volts, the cutout adjusting screw must be reset. Turn the screw clockwise to increase the reading or anti-clockwise to reduce it. Test after each adjustment and when the reading is correct, tighten the locknut. The cutout adjusting screw is equally sensitive and false readings are again liable to occur if adjustment is prolonged.

7   If the cutout points fail to close, there may be an open circuit in the regulator unit or in the dynamo itself. It is advisable to seek the assistance of a qualified auto-electrical expert at this stage.

8   Alternator: checking the output

1   The output and performance of the alternator fitted to the 1958 - 60 B31 and B33 can be checked only with specialised test equipment of the multi-meter type. It is unlikely that the average owner will have access to this type of equipment or instruction in its use. In consequence, if the performance is suspect, the alternator and charging circuit should be checked by a qualified auto-electrical expert.

2   Failure of the alternator does not necessarily mean that a replacement is needed. This can however sometimes be most economic through a service exchange scheme. It is possible to replace or rewind the stator coil assembly, for example, if the rotor is undamaged.

9   Selenium rectifier: general description

Alternator models only

1   The function of the selenium rectifier is to convert the AC produced by the alternator to DC so that it can be used to charge the battery and operate the lighting circuit etc. The usual symptom of a defective rectifier is a battery that discharges rapidly because it is receiving no charge from the generator.

2   The rectifier is located where it is not exposed to water or battery acid, which will cause it to malfunction. The question of access is of relatively little importance because the rectifier is unlikely to give trouble during normal operating conditions. It is not practicable to repair a damaged rectifier; replacement is the only satisfactory solution. One of the most frequent causes of rectifier failure is the inadvertent connection of the battery in reverse, which results in a reverse current flow.

3   It is not possible to check whether the rectifier is functioning correctly without the appropriate test equipment. A BSA agent or an auto-electrical expert are best qualified to advise in such cases.

4   Do not loosen the rectifier locking nut or bend, cut, scratch or rotate the selenium wafers. Any such action will cause the electrode alloy coating to peel and destroy the working action.

10   Battery: examination and maintenance

1   Several different types of Lucas lead-acid battery have been fitted to the BSA single cylinder models since their inception. Most batteries have a 13 amp hour capacity, but if the machine was specified for sidecar use, a 22 amp hour battery was fitted.

2   Battery maintenance is limited to keeping the electrolyte level just above the plates and separators.

3   Unless acid is spilt, which may occur if the machine falls over use only distilled water for topping up purposes, until the correct level is restored. If acid is spilt on any part of the machine, it should be neutralised immediately with an alkali such as washing soda or baking powder, and washed away with plenty of water. This will prevent corrosion from taking place. Top up in this

instance with sulphuric acid of the correct specific gravity (1.26 - 1.280).

4   It is seldom practicable to repair a cracked battery case because the acid that is already seeping through the crack will prevent the formation of an effective seal, no matter what sealing compound is used. It is always best to replace a cracked battery, especially in view of the risk of corrosion from the acid leakage.

5   Make sure the battery is clamped securely. A loose battery will vibrate and its working life will be greatly shortened, due to the paste being shaken out of the plates.

11   Battery: charging procedure

1   Whilst the machine is running, it is unlikely that the battery will require attention other than routine maintenance because the dynamo will keep it charged. However, if the machine is used for a succession of short journeys, mainly during the hours of darkness, it is possible that the dynamo will be unable to keep pace with the heavy electrical demand, especially on the earlier models that have a dynamo giving a lower output. Under these circumstances, it will be necessary to remove the battery from time to time, to have it charged independently.

2   The normal charging rate is 1 amp. A more rapid charge can be given in an emergency, but this should be avoided if possible because it will tend to shorten the useful life of the battery.

3   When the battery is removed from a machine that has been laid up, a 'refresher' charge should be given every six weeks, if the battery is to be maintained in good condition.

12   Headlamp: replacing bulbs and adjusting beam height

1   All models have an integral headlamp, in which the reflector and the headlamp glass are sealed together. This type of lamp rim is secured by a screw in the top of the headlamp shell, which must be slackened before the rim complete with lamp unit can be withdrawn. The main headlamp bulb is of the pre-focus type, which has a special type of bayonet-fitting connector. A rim that is integral with the bulb ensures the bulb can be replaced in only the correct position; earlier bulbs have the bayonet cap marked 'TOP' so that the main and dip contacts cannot be transposed inadvertently.

2   The pilot bulb holder is a push fit in the reflector or is fitted below the headlamp, having a separate lens. Both types of pilot bulb have a bayonet fitting.

3   Beam height is adjusted by slackening the two bolts that retain the headlamp shell to the forks and tilting the headlamp either upwards or downwards. Adjustments should always be made with the rider seated normally.

4   UK lighting regulations stipulate that the lighting system must be arranged so that the light does not dazzle a person standing in the same horizontal plane as the vehicle, at a distance greater than 25 yards from the lamp, whose eye level is not less than 3 feet 6 inches above that plane. It is easy to approximate this setting by placing the machine 25 yards away from a wall, on a level road, and setting the beam height so that it is concentrated at the same height as the distance from the centre of the headlamp to the ground. The rider must be seated normally during this operation, and the pillion passenger, if one is carried regularly.

5   If the headlamp bulb is broken, it can be removed from the headlamp rim by detaching the wire retaining clips, after the front has been removed from the headlamp. In the case of a headlamp of the unit glass/reflector type, it will be necessary to purchase the complete beam unit, and not the glass alone.

13   Tail and stop lamp: replacing bulbs

1   Although the very early models were supplied with a tail lamp only, it is doubtful whether any of the original fittings are still in use because the size no longer meets the minimum requirements of the lighting regulations. Most of the current tail

12.1a Headlamp rim is retained by a single screw

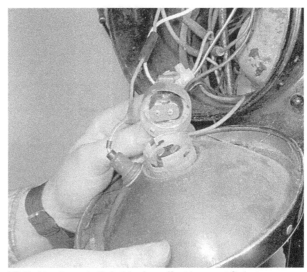

12.1b Headlamp bulb holder has offset bayonet fixing

12.1c Note locating cutout on bulb rim

12.2 Pilot bulb holder is a push fit in reflector

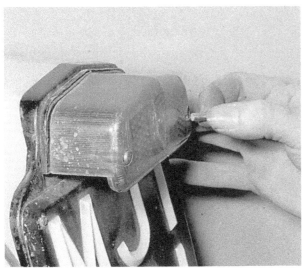

13.2a Tail/stop lamp lens cover is held by two screws

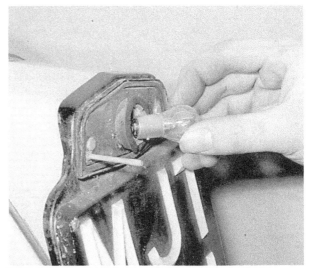

13.2b Bulb has offset pins to avoid incorrect replacement

lamp units contain provision for a stop lamp bulb also, which is operated when the rear brake pedal is depressed.

2   Removal of the plastics lens cover will reveal the bulb holders for both the tail lamp and the stop lamp, which may be separate, as in the case of the early models or combined, to conform to current practice. It is now customary to fit a single bulb with offset pins, which is of the double filament type. The offset pins prevent accidental inversion of the bulb. The tail lamp filament is rated at 6W and the stop lamp filament at 18W.

3   The stop lamp switch will be found on the left-hand side of the machine, in close proximity to the brake pedal. The switch does not require attention other than the occasional drop of light oil.

14 Wiring: layout and examination

1   The wiring is colour-coded and will correspond with the accompanying wiring diagrams.

2   Visual inspection will show whether any breaks or frayed outer coverings are giving rise to short circuits. Another source of trouble may be the snap connectors, particularly where the connector has not been pushed home fully in the outer casing. Early models are especially prone to wiring faults because the rubber-covered cables used at that period deteriorate as time progresses.

3   Intermittent short circuits can sometimes be traced to a chafed wire that passes through a frame member. Avoid tight bends in the wire or situations where the wire can become trapped or stretched between casings.

15 Headlamp switch

1   It is unusual for the headlamp switch to give trouble unless the machine has been laid up for a considerable period and the switch contacts have become dirty. Contact between the terminal posts is made by a spring-loaded roller attached to the body of the switch knob. If the terminal posts become corroded or oxidised, poor or intermittent electrical contact will result.

2   It is possible to dismantle the switch and clean the terminal posts and roller by hand; the rotor containing the roller can be pulled away when the centre screw of the switch knob is removed and the knob detached. This is a somewhat delicate operation, which should be performed only when the switch complete is removed from the headlamp shell. The switch body is held to the underside of the shell by means of a wire spring that engages with a groove around the body moulding.

3   A better alternative that does not necessitate dismantling the switch is the use of one of the proprietary switch contact cleansers that are available in aerosol form.

4   On no account oil the switch or oil will spread across the internal contacts, to form an effective insulator.

16 Ignition switch (alternator models)

1   The alternator on B31 and B33 models is fitted with a combined lighting/ignition switch similar in principle to the switch fitted to the Magdyno models, but which accepts an ignition key.

2   Remarks made in the previous Section concerning the cleaning of the switch apply equally.

3   In addition to the ignition on position, a position is given for starting the machine in the event of a flat battery. When placed in the Emergency position, current is allowed to flow directly from the alternator to the ignition coil. Unfortunately the system is not absolutely reliable, and even with an alternator and electrical system in perfect condition, the Emergency method of starting is not always effective.

17 Fault diagnosis: electrical system

Symptom	Cause	Remedy
Complete electrical failure	Isolated battery	Check battery connections, also whether connections show signs of corrosion.
Dim lights, horn inoperative	Discharged battery	Remove battery and charge with battery charger. Check generator output and/or voltage regulator settings.
Constantly blowing bulbs	Vibration or poor earth connection	Check security of bulb holders. Check earth return connections.
Parking lights dim rapidly	Battery will not hold charge	Renew battery at earliest opportunity.
Tail lamp fails	Blown bulb	Renew.
Headlamp fails	Blown bulb	Renew.
Horn inoperative or weak	Faulty switch Out of adjustment	Check switch. Re-adjust.
Incorrect charging	Faulty rectifier (alternator models) Faulty regulator (dynamo models) Wiring fault	Check. Check and adjust. Check.
Over or under-charging	As above, or faulty battery	Check.

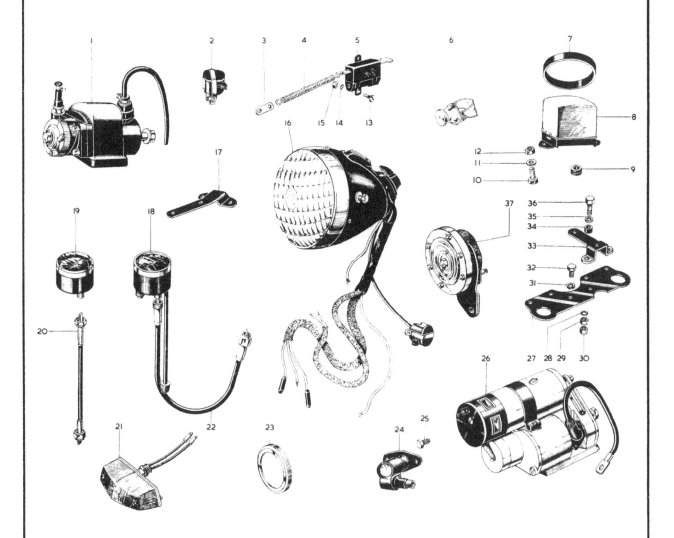

Fig. 8.5. Electrical equipment - component parts

1	Magneto (competition models)	10	Bolt - 2 off	20	Revolution counter cable	29	Nut - 4 off
2	Dip switch	11	Washer - 2 off	21	Stop-tail lamp unit	30	Locknut - 4 off
3	Stop lamp switch link	12	Nut - 2 off	22	Speedometer cable	31	Washer - 2 off
4	Stop lamp switch spring	13	Screw - 2 off	23	Licence holder	32	Screw - 2 off
5	Stop lamp switch unit	14	Washer - 2 off	24	Revolution counter drive gearbox	33	Speedometer bracket
6	Horn push	15	Nut - 2 off			34	Isolating rubber
7	Control box rubber buffer	16	Headlamp unit	25	Bolt - 2 off	35	Washer - 4 off
8	Voltage control box	17	Mounting bracket	26	Magdyno (road models)	36	Bolt - 2 or 4 off
9	Grommet	18	Speedometer head	27	Instrument mounting plate	37	Horn
		19	Revolution counter	28	Washer - 4 off		

138

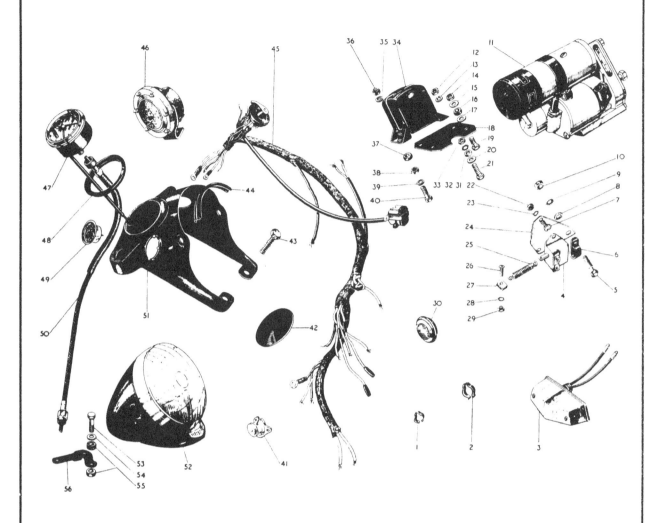

Fig. 8.6. Electrical equipment - component parts

1 Cable clip (2 off)
2 Cable clip (2 off)
3 Stop-tail lamp
4 Stop lamp switch
5 Bolt
6 Clip
7 Screw (2 off)
8 Washer
9 Washer
10 Nut
11 Magdyno
12 Nut (2 off)
13 Spring washer (2 off)
14 Nut (2 off)

15 Washer (2 off)
16 Rubber washer (2 off)
17 Washer (2 off)
18 Mounting bracket
19 Bolt (2 off)
20 Washer (2 off)
21 Bolt (2 off)
22 Nut (2 off)
23 Washer (2 off)
24 Switch plate
25 Spring
26 Screw
27 Clip
28 Washer

29 Nut
30 Rear reflector
31 Washer (2 off)
32 Washer (2 off)
33 Nut (2 off)
34 Voltage regulator
35 Washer (2 off)
36 Nut (2 off)
37 Rubber washer (2 off)
38 Locknut (2 off)
39 Washer (2 off)
40 Bolt (2 off)
41 Horn push
42 Cowl blanking plug

43 Bolt (2 off)
44 Rubber bead
45 Harness, light and dip switch
46 Horn
47 Speedometer head
48 Rubber mounting ring
49 Ammeter
50 Speedometer cable
51 Headlamp cowl
52 Headlamp/sidelight unit
53 Bolt (2 off)
54 Washer (2 off)
55 Rubber washer (2 off)
56 Mounting bracket (2 off)

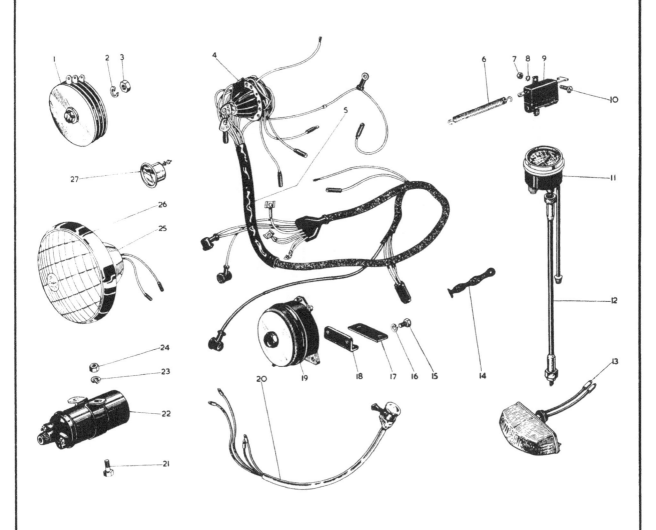

Fig. 8.7. Electrical equipment, coil ignition models - component parts

1	Rectifier	8	Washer (2 off)	15	Bolt (2 off)	22	Ignition coil
2	Spring washer	9	Stop lamp switch	16	Spring washer (2 off)	23	Spring washer (2 off)
3	Nut	10	Screw (2 off)	17	Plate	24	Nut (2 off)
4	Lighting switch	11	Speedometer head	18	Bracket	25	Headlamp unit
5	Electrical harness	12	Speedometer cable	19	Horn	26	Headlamp rim
6	Stop lamp switch spring	13	Stop-tail lamp	20	Horn/clip switch	27	Ammeter
7	Nut (2 off)	14	Wiring clip (15 off)	21	Bolt		

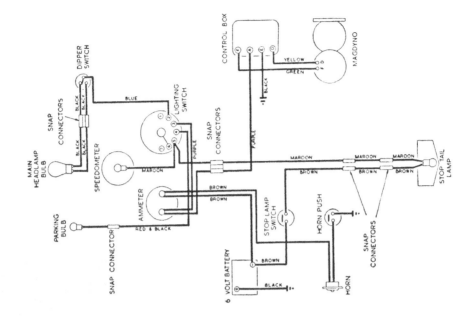

Fig. 8.8. 1954-1955 swinging arm models. Plunger and rigid frame models similar except horn connected to negative battery

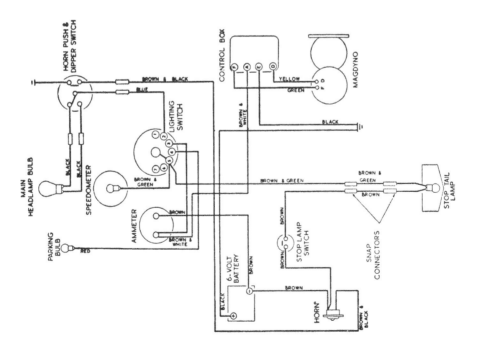

Fig. 8.9. 1956-7 swinging arm models and 1956-61 plunger suspension models. On swinging arm models horn is connected to ammeter as in 1954-1955 wiring diagram.

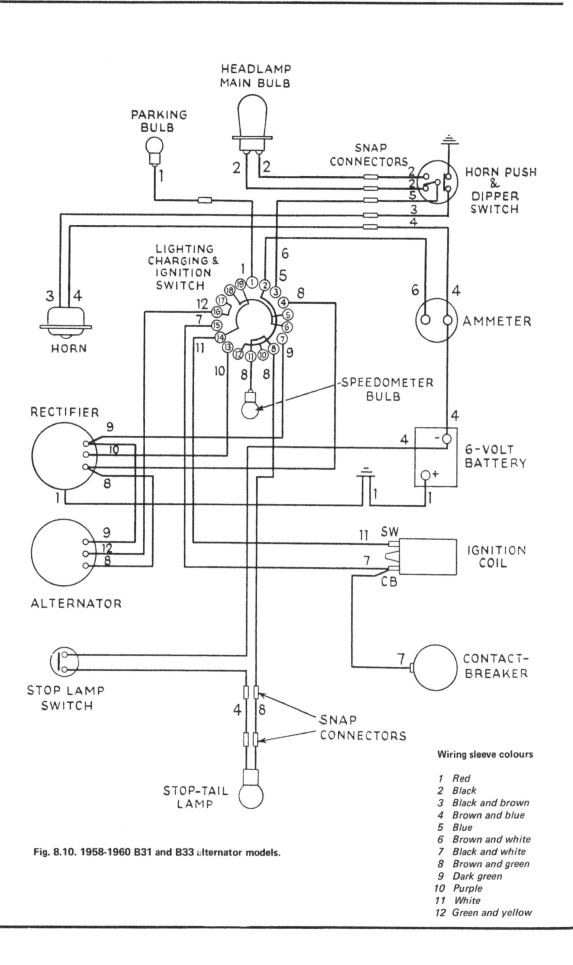

Fig. 8.10. 1958-1960 B31 and B33 alternator models.

Wiring sleeve colours

1 Red
2 Black
3 Black and brown
4 Brown and blue
5 Blue
6 Brown and white
7 Black and white
8 Brown and green
9 Dark green
10 Purple
11 White
12 Green and yellow

Conversion Factors

Length (distance)

Inches (in)	x 25.4	= Millimetres (mm)	x 0.0394	= Inches (in)	
Feet (ft)	x 0.305	= Metres (m)	x 3.281	= Feet (ft)	
Miles	x 1.609	= Kilometres (km)	x 0.621	= Miles	

Volume (capacity)

Cubic inches (cu in; in³)	x 16.387	= Cubic centimetres (cc; cm³)	x 0.061	= Cubic inches (cu in; in³)	
Imperial pints (Imp pt)	x 0.568	= Litres (l)	x 1.76	= Imperial pints (Imp pt)	
Imperial quarts (Imp qt)	x 1.137	= Litres (l)	x 0.88	= Imperial quarts (Imp qt)	
Imperial quarts (Imp qt)	x 1.201	= US quarts (US qt)	x 0.833	= Imperial quarts (Imp qt)	
US quarts (US qt)	x 0.946	= Litres (l)	x 1.057	= US quarts (US qt)	
Imperial gallons (Imp gal)	x 4.546	= Litres (l)	x 0.22	= Imperial gallons (Imp gal)	
Imperial gallons (Imp gal)	x 1.201	= US gallons (US gal)	x 0.833	= Imperial gallons (Imp gal)	
US gallons (US gal)	x 3.785	= Litres (l)	x 0.264	= US gallons (US gal)	

Mass (weight)

Ounces (oz)	x 28.35	= Grams (g)	x 0.035	= Ounces (oz)	
Pounds (lb)	x 0.454	= Kilograms (kg)	x 2.205	= Pounds (lb)	

Force

Ounces-force (ozf; oz)	x 0.278	= Newtons (N)	x 3.6	= Ounces-force (ozf; oz)	
Pounds-force (lbf; lb)	x 4.448	= Newtons (N)	x 0.225	= Pounds-force (lbf; lb)	
Newtons (N)	x 0.1	= Kilograms-force (kgf; kg)	x 9.81	= Newtons (N)	

Pressure

Pounds-force per square inch (psi; lbf/in²; lb/in²)	x 0.070	= Kilograms-force per square centimetre (kgf/cm²; kg/cm²)	x 14.223	= Pounds-force per square inch (psi; lbf/in²; lb/in²)
Pounds-force per square inch (psi; lbf/in²; lb/in²)	x 0.068	= Atmospheres (atm)	x 14.696	= Pounds-force per square inch (psi; lbf/in²; lb/in²)
Pounds-force per square inch (psi; lbf/in²; lb/in²)	x 0.069	= Bars	x 14.5	= Pounds-force per square inch (psi; lbf/in²; lb/in²)
Pounds-force per square inch (psi; lbf/in²; lb/in²)	x 6.895	= Kilopascals (kPa)	x 0.145	= Pounds-force per square inch (psi; lbf/in²; lb/in²)
Kilopascals (kPa)	x 0.01	= Kilograms-force per square centimetre (kgf/cm²; kg/cm²)	x 98.1	= Kilopascals (kPa)
Millibar (mbar)	x 100	= Pascals (Pa)	x 0.01	= Millibar (mbar)
Millibar (mbar)	x 0.0145	= Pounds-force per square inch (psi; lbf/in²; lb/in²)	x 68.947	= Millibar (mbar)
Millibar (mbar)	x 0.75	= Millimetres of mercury (mmHg)	x 1.333	= Millibar (mbar)
Millibar (mbar)	x 0.401	= Inches of water (inH₂O)	x 2.491	= Millibar (mbar)
Millimetres of mercury (mmHg)	x 0.535	= Inches of water (inH₂O)	x 1.868	= Millimetres of mercury (mmHg)
Inches of water (inH₂O)	x 0.036	= Pounds-force per square inch (psi; lbf/in²; lb/in²)	x 27.68	= Inches of water (inH₂O)

Torque (moment of force)

Pounds-force inches (lbf in; lb in)	x 1.152	= Kilograms-force centimetre (kgf cm; kg cm)	x 0.868	= Pounds-force inches (lbf in; lb in)
Pounds-force inches (lbf in; lb in)	x 0.113	= Newton metres (Nm)	x 8.85	= Pounds-force inches (lbf in; lb in)
Pounds-force inches (lbf in; lb in)	x 0.083	= Pounds-force feet (lbf ft; lb ft)	x 12	= Pounds-force inches (lbf in; lb in)
Pounds-force feet (lbf ft; lb ft)	x 0.138	= Kilograms-force metres (kgf m; kg m)	x 7.233	= Pounds-force feet (lbf ft; lb ft)
Pounds-force feet (lbf ft; lb ft)	x 1.356	= Newton metres (Nm)	x 0.738	= Pounds-force feet (lbf ft; lb ft)
Newton metres (Nm)	x 0.102	= Kilograms-force metres (kgf m; kg m)	x 9.804	= Newton metres (Nm)

Power

Horsepower (hp)	x 745.7	= Watts (W)	x 0.0013	= Horsepower (hp)

Velocity (speed)

Miles per hour (miles/hr; mph)	x 1.609	= Kilometres per hour (km/hr; kph)	x 0.621	= Miles per hour (miles/hr; mph)

Fuel consumption*

Miles per gallon (mpg)	x 0.354	= Kilometres per litre (km/l)	x 2.825	= Miles per gallon (mpg)

Temperature

Degrees Fahrenheit = (°C x 1.8) + 32   Degrees Celsius (Degrees Centigrade; °C) = (°F - 32) x 0.56

It is common practice to convert from miles per gallon (mpg) to litres/100 kilometres (l/100km), where mpg x l/100 km = 282

Safety first!

Professional motor mechanics are trained in safe working procedures. However enthusiastic you may be about getting on with the job in hand, do take the time to ensure that your safety is not put at risk. A moment's lack of attention can result in an accident, as can failure to observe certain elementary precautions.

There will always be new ways of having accidents, and the following points do not pretend to be a comprehensive list of all dangers; they are intended rather to make you aware of the risks and to encourage a safety-conscious approach to all work you carry out on your vehicle.

Essential DOs and DON'Ts

DON'T start the engine without first ascertaining that the transmission is in neutral.

DON'T suddenly remove the filler cap from a hot cooling system – cover it with a cloth and release the pressure gradually first, or you may get scalded by escaping coolant.

DON'T attempt to drain oil until you are sure it has cooled sufficiently to avoid scalding you.

DON'T grasp any part of the engine, exhaust or silencer without first ascertaining that it is sufficiently cool to avoid burning you.

DON'T allow brake fluid or antifreeze to contact the machine's paintwork or plastic components.

DON'T syphon toxic liquids such as fuel, brake fluid or antifreeze by mouth, or allow them to remain on your skin.

DON'T inhale dust – it may be injurious to health (see *Asbestos* heading).

DON'T allow any spilt oil or grease to remain on the floor – wipe it up straight away, before someone slips on it.

DON'T use ill-fitting spanners or other tools which may slip and cause injury.

DON'T attempt to lift a heavy component which may be beyond your capability – get assistance.

DON'T rush to finish a job, or take unverified short cuts.

DON'T allow children or animals in or around an unattended vehicle.

DON'T inflate a tyre to a pressure above the recommended maximum. Apart from overstressing the carcase and wheel rim, in extreme cases the tyre may blow off forcibly.

DO ensure that the machine is supported securely at all times. This is especially important when the machine is blocked up to aid wheel or fork removal.

DO take care when attempting to slacken a stubborn nut or bolt. It is generally better to pull on a spanner, rather than push, so that if slippage occurs you fall away from the machine rather than on to it.

DO wear eye protection when using power tools such as drill, sander, bench grinder etc.

DO use a barrier cream on your hands prior to undertaking dirty jobs – it will protect your skin from infection as well as making the dirt easier to remove afterwards; but make sure your hands aren't left slippery. Note that long-term contact with used engine oil can be a health hazard.

DO keep loose clothing (cuffs, tie etc) and long hair well out of the way of moving mechanical parts.

DO remove rings, wristwatch etc, before working on the vehicle – especially the electrical system.

DO keep your work area tidy – it is only too easy to fall over articles left lying around.

DO exercise caution when compressing springs for removal or installation. Ensure that the tension is applied and released in a controlled manner, using suitable tools which preclude the possibility of the spring escaping violently.

DO ensure that any lifting tackle used has a safe working load rating adequate for the job.

DO get someone to check periodically that all is well, when working alone on the vehicle.

DO carry out work in a logical sequence and check that everything is correctly assembled and tightened afterwards.

DO remember that your vehicle's safety affects that of yourself and others. If in doubt on any point, get specialist advice.

IF, in spite of following these precautions, you are unfortunate enough to injure yourself, seek medical attention as soon as possible.

Asbestos

Certain friction, insulating, sealing, and other products – such as brake linings, clutch linings, gaskets, etc – contain asbestos. *Extreme care must be taken to avoid inhalation of dust from such products since it is hazardous to health.* If in doubt, assume that they *do* contain asbestos.

Fire

Remember at all times that petrol (gasoline) is highly flammable. Never smoke, or have any kind of naked flame around, when working on the vehicle. But the risk does not end there – a spark caused by an electrical short-circuit, by two metal surfaces contacting each other, by careless use of tools, or even by static electricity built up in your body under certain conditions, can ignite petrol vapour, which in a confined space is highly explosive.

Always disconnect the battery earth (ground) terminal before working on any part of the fuel or electrical system, and never risk spilling fuel on to a hot engine or exhaust.

It is recommended that a fire extinguisher of a type suitable for fuel and electrical fires is kept handy in the garage or workplace at all times. Never try to extinguish a fuel or electrical fire with water.

Note: *Any reference to a 'torch' appearing in this manual should always be taken to mean a hand-held battery-operated electric lamp or flashlight. It does **not** mean a welding/gas torch or blowlamp.*

Fumes

Certain fumes are highly toxic and can quickly cause unconsciousness and even death if inhaled to any extent. Petrol (gasoline) vapour comes into this category, as do the vapours from certain solvents such as trichloroethylene. Any draining or pouring of such volatile fluids should be done in a well ventilated area.

When using cleaning fluids and solvents, read the instructions carefully. Never use materials from unmarked containers – they may give off poisonous vapours.

Never run the engine of a motor vehicle in an enclosed space such as a garage. Exhaust fumes contain carbon monoxide which is extremely poisonous; if you need to run the engine, always do so in the open air or at least have the rear of the vehicle outside the workplace.

The battery

Never cause a spark, or allow a naked light, near the vehicle's battery. It will normally be giving off a certain amount of hydrogen gas, which is highly explosive.

Always disconnect the battery earth (ground) terminal before working on the fuel or electrical systems.

If possible, loosen the filler plugs or cover when charging the battery from an external source. Do not charge at an excessive rate or the battery may burst.

Take care when topping up and when carrying the battery. The acid electrolyte, even when diluted, is very corrosive and should not be allowed to contact the eyes or skin.

If you ever need to prepare electrolyte yourself, always add the acid slowly to the water, and never the other way round. Protect against splashes by wearing rubber gloves and goggles.

Mains electricity and electrical equipment

When using an electric power tool, inspection light etc, always ensure that the appliance is correctly connected to its plug and that, where necessary, it is properly earthed (grounded). Do not use such appliances in damp conditions and, again, beware of creating a spark or applying excessive heat in the vicinity of fuel or fuel vapour. Also ensure that the appliances meet the relevant national safety standards.

Ignition HT voltage

A severe electric shock can result from touching certain parts of the ignition system, such as the HT leads, when the engine is running or being cranked, particularly if components are damp or the insulation is defective. Where an electronic ignition system is fitted, the HT voltage is much higher and could prove fatal.

English/American terminology

Because this book has been written in England, British English component names, phrases and spellings have been used throughout. American English usage is quite often different and whereas normally no confusion should occur, a list of equivalent terminology is given below.

English	American	English	American
Air filter	Air cleaner	Number plate	License plate
Alignment (headlamp)	Aim	Output or layshaft	Countershaft
Allen screw/key	Socket screw/wrench	Panniers	Side cases
Anticlockwise	Counterclockwise	Paraffin	Kerosene
Bottom/top gear	Low/high gear	Petrol	Gasoline
Bottom/top yoke	Bottom/top triple clamp	Petrol/fuel tank	Gas tank
Bush	Bushing	Pinking	Pinging
Carburettor	Carburetor	Rear suspension unit	Rear shock absorber
Catch	Latch	Rocker cover	Valve cover
Circlip	Snap ring	Selector	Shifter
Clutch drum	Clutch housing	Self-locking pliers	Vise-grips
Dip switch	Dimmer switch	Side or parking lamp	Parking or auxiliary light
Disulphide	Disulfide	Side or prop stand	Kick stand
Dynamo	DC generator	Silencer	Muffler
Earth	Ground	Spanner	Wrench
End float	End play	Split pin	Cotter pin
Engineer's blue	Machinist's dye	Stanchion	Tube
Exhaust pipe	Header	Sulphuric	Sulfuric
Fault diagnosis	Trouble shooting	Sump	Oil pan
Float chamber	Float bowl	Swinging arm	Swingarm
Footrest	Footpeg	Tab washer	Lock washer
Fuel/petrol tap	Petcock	Top box	Trunk
Gaiter	Boot	Torch	Flashlight
Gearbox	Transmission	Two/four stroke	Two/four cycle
Gearchange	Shift	Tyre	Tire
Gudgeon pin	Wrist/piston pin	Valve collar	Valve retainer
Indicator	Turn signal	Valve collets	Valve cotters
Inlet	Intake	Vice	Vise
Input shaft or mainshaft	Mainshaft	Wheel spindle	Axle
Kickstart	Kickstarter	White spirit	Stoddard solvent
Lower leg	Slider	Windscreen	Windshield
Mudguard	Fender		

Index

A

Adjustments
  brakes - 125
  brake cables - 11
  carburettor - 89
  clutch - 11, 78
  contact breaker - 95
  drive gear - 97
  primary chaing 9, 74
  rear suspension - 112
Air filter - 10, 89
Alternator - 55, 134
Auto-advance unit - 98

B

Badges - 114
Battery - 7, 134
Bearings, wheel - 120, 125
Brakes
  adjustment - 125
  drum removal - 125
  examination and replacement - 118, 125
  fault diagnosis - 128
  specifications - 116
Bulbs
  headlamp - 134
  tail and stop lamp - 134

C

Camshaft - 32, 33, 44
Capacities - 80
Carburettor
  adjustment - 89
  dismantling - 84
  fault diagnosis - 91
  removal - 84
  specifications - 80
Centre stand - 112
Chain, final drive - 126
  rear - 79
Cleaning - 115
Clutch
  adjustment - 78
  dismantling - 75
  examination and renovation - 75
  fault diagnosis - 79
  reassembly - 78
  removal - 18
  specification - 75
Coil - 98
Condenser - 96, 98
Contact breaker
  adjustment - 95, 97
  examination and renovation - 95
  refitting - 49
  removal - 32, 95, 98
Crankcase - 34, 42, 44
Cylinder barrel - 24, 38, 46
Cylinder head - 24, 40, 49

D

Drive gears - 32
Dual seat - 114
Dynamo - 131

E

Electrical system
  fault diagnosis - 136
  specifications - 130
Engine
  dismantling - 18, 24
  examination and renovation - 37
  fault diagnosis - 56
  lubrication - 89
  operations - 17
  reassembly - 42
  removal - 18, 20
  specifications - 14
  starting and running - 56
Exhaust system - 89

F

Fault diagnosis
  clutch - 79
  electrical system - 136
  engine - 56
  frame and forks - 115
  gearbox - 7
  fuel system - 91
  gearbox - 74
  ignition system - 100
  lighting - 136
  lubrication - 91
  wheels, brakes and tyres - 128
Flywheel - 37
Final drive chain - 126
Footrests - 114
Frame
  examination and renovation - 108
  fault diagnosis - 115
  specifications - 101
Front forks
  damping - 108
  dismantling - 104
  examination - 106
  fault diagnosis - 115
  reassembly - 107
  removal - 101
  specifications - 101
Fuel system
  fault diagnosis - 91
  specifications - 80

G

Gearbox
  dismantling - 60
  examination and renovation - 65
  fault diagnosis - 74

oil level - 9
reassembly - 69
specifications - 58

H

Headlamp - 134, 136

I

Ignition
    coil - 98
    condenser - 96, 98
    contact breaker - 95
    fault diagnosis - 100
    magdyno - 31, 46, 93
    magneto - 31, 46, 94
    specifications - 91
    switch - 136
    timing - 49, 97, 98

K

Kickstart - 67, 71

L

Lights - 134
Lubricants, recommended - 12
Lubrication system - 9, 11, 89, 90

M

Magdyno - 31, 46
Magneto - 31, 46, 97
Main bearings - 37
Maintenance, routine - 7
Modifications - 5

O

Oil filters - 90
Oil pump - 34, 44, 90
Oil seals - 106
Ordering spare parts - 6

P

Petrol pipes - 83
Petrol tank - 18, 83
Petrol tap - 83
Piston - 31, 39, 46
Piston rings - 39
Primary chaincase - 9, 74
Primary drive - 18
Prop stand - 112

R

Rear chain - 9
Rear suspension units - 112
Rear wheel sprocket - 126
Recommended lubricants - 12
Rectifier - 134

Regulator - 132
Rockers - 42
Routine maintenance - 7

S

Selenium rectifier - 134
Small end bush - 38
Spare parts, ordering - 6
Spark plugs - 100
Specifications
    brakes - 116
    clutch - 75
    electrical system - 130
    engine - 14
    frame and forks - 101
    fuel system - 80
    ignition system - 92
    lubrication system - 80
    tyres - 116
    wheels - 116
Speedometer - 114
Steering damper - 114
Steering head bearings - 107
Steering head lock - 114
Stop lamp - 134
Swinging arm rear suspension - 108

T

Tachometer - 114
Tail lamp - 134
Tank badges - 114
Timing, ignition - 49, 97, 98
Tools - 12
Tyres
    pressures - 7
    removal and replacement - 126
    specifications - 116
    vaive dust caps - 128

V

Valves - 10, 39, 51
Voltage regulator - 132

W

Wheels
    balance - 126
    bearing - 120
    fault diagnosis - 128
    front - 116, 122
    rear - 122, 126
    replacement - 126
    specifications - 116
Wiring diagrams - 140
    layout and inspection - 136
Working conditions - 12